Technology's Storytellers

Technology's Storytellers

Reweaving the Human Fabric

John M. Staudenmaier, S.J.

Published jointly by
The Society for the History of Technology
and
The MIT Press
Cambridge, Massachusetts
London, England

This book was set in Baskerville by The MIT Press Computergraphics Department and printed and bound by The Murray Printing Company in the United States of America.

Library of Congress Cataloging in Publication Data

Staudenmaier, John M.
Technology's storytellers.

Bibliography: p.
Includes index.
1. Technology—History. 2. Technology—Social aspects. I. Title.
T20.S73 1984 306'.46 84-23412
ISBN 0-262-19237-3

To my mother and father, and to the Lakota people of Pine Ridge, South Dakota

Contents

List of Tables

Foreword

This volume represents a happy convergence between this careful study of the first twenty years of the Society for the History of Technology and the desire of the society to recognize Melvin Kranzberg, the editor during those seminal years. Kranzberg was the principal founder of the society and, as first editor of its journal, was largely responsible for its character and direction of motion. Moreover, as first secretary of the society, he made most of the moves and decisions related to those foundations from which articles and books would emerge.

As editor, his greatest success was in reaching out to encourage people from many backgrounds to prepare and submit papers to *Technology and Culture*. In every possible way, he expanded the reach of the journal in order to justify its title. He wanted it to be international in scope and of as much importance for general historians as for historically minded engineers.

He started with a motley crew of scholars who were but slightly conscious of their common interest in the history of technology. Individually, they thought of themselves first as engineers, business historians, economists, sociologists, historians of science, or general historians who had wandered into a somewhat strange field. By the time Kranzberg withdrew from the editorship, the history of technology had been professionalized with its own educational and training patterns and its own standards of rigor; it had become a community of scholars.

Kranzberg sought never to cut down anyone who had a possible interest in the history of technology. Even when the person in question had inadequate understanding and wrote unacceptably, he tried to turn the author toward improving his work—not to discourage him. Even if a submitted paper was manifestly alien to the field, he would indicate needed improvement and point to a more appropriate journal.

He put together the best available panel of editorial advisors and relied fully on their reactions. He respected them and their knowledge—just as he respected the best efforts of those who submitted papers.

By recognizing an incipient new field, Kranzberg and those who worked with him in the early years brought the field into bloom. But like all inventors and innovators within the history of technology, he was bound by the state of the art. He might encourage new and expanded inquiries, but he could only accept the dimensions of the field for what they were at any given moment. The spectrum of existing knowledge, analysis, and synthesis kept changing, but *Technology and Culture* could reflect nothing more than what was emerging from the study of the history of technology as each number appeared. As a result, what the field was from point to point and how it changed is better reflected in *Technology and Culture* than in any other source.

But what was it? How is one to interpret the varied and often unrelated contributions? They might, of course, be catalogued and numbered. They could even be evaluated in terms of quality or influence, but the result would have to be somewhat sterile and short of the need. A conventional festschrift might have brought together experts in various subfields to explore varied dimensions of the world of Kranzberg's *Technology and Culture*. That would not have been enough; a more focused and more interpretive study was called for.

Technology's Storytellers provides that. John M. Staudenmaier has read the contributions to *Technology and Culture* through these years with a precision of purpose no one else has attempted. He has thought about each in terms of his own excellent preparation in the larger study of the field and in terms of his own background, which he carefully explains. He has interpreted the whole in a masterly manner.

The result is a creative study in its own right. It examines the course of *Technology and Culture* from the perspective of the current professional study of the history of technology. It reviews the field from the outlook of most emerging scholars of today. Because it represents a sensitive interpretation of very recent history, it will stimulate just the sort of discourse Staudenmaier has correctly identified as a characteristic of the present flourishing field. It will be an important resource.

It is appropriate that this study should be published by the Society for the History of Technology and the MIT Press, with the specific aid of Bern Dibner, a longtime supporter of the field. The two institutions collaborated in an important Monograph Series that produced several books important to the field, some of them milestones. The present work represents a renewal of this collaboration and may predict a renewal of the series, as it points so clearly to the future in many other ways.

Brooke Hindle

Introduction

In every society the storyteller wrestles with the challenge and artistry of the historical endeavor. Whether the tale is told with the help of computerized data analysis or by means of delicate sketches on a Plains Indian winter count robe, the historian's account of the past is essential to the life of the culture. The stories create a context of origin, that the people may not live alienated from their ancestry and in ignorance of the events that have given shape to their present. By naming the central actors and events of the culture's heritage and by interpreting their meaning, the historian crafts a historically based language, which is the basis for the self-knowledge of any people.

In our present circumstances—I refer to the late twentieth century and to those cultures generally called "the West"—the profound influence of technology poses a challenge to the historical endeavor. It is a commonplace observation that increasingly complex technological networks have dramatically changed the cultural world of the West. Equally important, the extraordinary westernization of technologies in other societies has created a cultural problem involving the entire planet. For both reasons the contribution of the historian of technology is critical. How will we, in the West, tell the tale of our technological past? What interpretative language will shape our frame of reference for thinking about and responding to current technological issues? How will we understand the relationship between Western technology and nonwestern societies?

These are not idle questions. The technological mind-set and self-knowledge of a people so strongly influenced by technology not only will govern the integration of technical decisions within our larger culture but will shape our relationships with other societies in terms of technological exchange. These are the questions that led me to the work you are about to read.

Starting points are significant for any interpretation, including this study. While they limit our perspectives, they also give us eyes to see and generate a set of questions to ask in research. Because every interpretation begins with some starting point and its attendant biases, it is important to claim my own and to explain how the question of "design-ambient integration" became so important in this study. I came to the history of technology after nearly a decade spent living and teaching on the Pine Ridge Reservation of the Oglala Lakota (Sioux) in South Dakota. When I arrived there in 1964, fresh from a master's degree in philosophy (St. Louis University, thesis on Bergson), I was a citizen of the West, a Cartesian raationalist who believed implicitly in the ordinary canons of knowledge respected by my culture's intellectual tradition. Gradually, during the years on Pine Ridge, I found my rationalist world broken open by a number of transforming experiences. One was the slow recognition that I lived within a culture radically different from my own. I began to realize that Lakota people interpret the universe from a different perspective. They are contemplative more than active; their experience of time is centered in events rather than in an abstract clock system; they experience themselves as kin and not conquerors of the rest of natural reality; and they honor generosity and communal sharing more than individual initiative and personal success. At the same time I experienced suffering at a level I had not known. Extreme poverty—Shannon County, the heart of Pine Ridge, was the poorest county in the entire United States in the 1980 census—is not attractive at close range. It includes sickness and oppression, anger and grief. At the heart of that experience of suffering was, I began to believe, a profound temptation to despair, a temptation that was and is embodied in shocking levels of suicide, drinking, and sudden death. I have not been able to make an accurate count, but I suspect that between thirty and forty of the students from the four years I taught at Pine Ridge are now dead. Finally, in the midst of this radically different culture and its despair, I found myself gradually being welcomed and loved by a people whose capacity for forgiveness, for humor, for loving kinship continues to move me to this day.

It is from this experience that the question of design-ambient integration emerged. Sometime in the last years of the sixties I began to ask myself where all this despair comes from. I began to answer my question—tentatively—by suggesting that the despair was due in great part to the bad fit between the values of the Lakota people and the values embedded in the technologies of the West that had become a part of life on Pine Ridge. To participate in the Western technological world, the Lakota had to abandon event-centered time for clock time, to adopt habits that fostered activism and caused contemplation to atrophy, to learn to act in an aggressive and individualistic manner rather than a communal one. I began to suspect that if this was happening here then it was very likely happening everywhere on earth where small and delicate cultures were being invaded by Western technologies. I found myself pondering such things as the role of the small transistor radio as a culture hammer.

During all this thinking in the late sixties I also came to realize that I knew very little about what "technology" was. Eventually, in 1973, I returned to study at the University of Pennsylvania in Philadelphia to educate myself in my areas of ignorance. There I encountered the young field of the history of technology in the Department of the History and Sociology of Science and in the person of Thomas P. Hughes, who gradually became a friend and a mentor. As the years of study passed, I found that my question about the source of despair on Pine Ridge changed in one significant way. I began to ask not whether "technology" caused despair but whether an ideology of technology was the source of the trouble. I began to focus my attention on the language of technological "progress." The myth of progress, of a singular posthistorical force sometimes called Science, sometimes Technology, sometimes both, began to appear as a pervasive belief system in the West. I came to realize that I had already met this myth in the racist jokes and the patronizing cynicism that frequently came my way when I introduced myself as someone living and teaching with Native Americans. I had long since realized that many of my middle-class American kin perceived Indians as a backward race that had mysteriously not been able to adapt to modern life, a people who were lazy and who spent their time drinking rather than achieving anything of worth.

The more I pursued the myth of progress, the more it seemed to be an ideology that fostered both passivity and violent aggression among its own constituents. The passivity was aptly expressed in the motto of the 1933 Chicago "Century of Progress' International Exposition: "Science Finds, Industry Applies, Man Conforms." Since Science and Technology progressed in an autonomous forward direction, no matter what one did about it, the only remaining role for the human person was to conform. At the same time, since this progressive force originated in the West and continued to be centered there, it came to be the ideological justification for Western colonialism of the past two centuries. It was the destiny of the West to be the cutting edge of human progress. This destiny required the combined exploitation and paternalistic "saving" of the backward cultures of the world. In the name of cultural superiority centuries of violence, of which Pine Ridge was a microcosm, have been seen as inevitable and even beneficial.

It is not difficult to see from this background story that I came to my study with a personal bias. If past technological change is commonly discussed as if it were independent of any historical context, then the language used to speak of present technological issues will be radically impoverished. For this reason I find the ahistorical nature of much popular technological analysis alarming. By the same token, I am concerned that so much general historical discourse pays so little attention to technology.

Human language is culture-bound and limited, but its power is immense, particularly in its inherent tendency to create the boundaries of our thought and imagination. To be attentive to how we talk is to become intimate with the possibilities and the limitations of our imaginations. My sense of the power of the technological language we have available to us, of the stories we tell one another about our past and ourselves, was the most important factor in my approach to this study. What follows is a textual analysis that will assess the value of a scholarly, historical approach in creating a new language for talking about technology, an approach that is free from the imaginative constraints of the "progress talk" that I have come to find so dangerous in the West. It is this concern about language that led me to spend so much time with the small group of scholarly storytellers who are the subject of this study.

A shared conviction about the importance of historical research into the technological past led a handful of scholars to form the Society for the History of Technology (SHOT) in 1958 and to begin publishing a quarterly journal, *Technology and Culture* (TC) in 1959. In the two decades since its birth, the society and its journal have provided a forum for an increasingly vigorous new discipline. Through the communal process of scholarly discourse, SHOT historians have begun to achieve some consensus about the methodological canons of the field and the critical thematic questions central to it.

My purpose in conducting this study has been to understand the emerging intellectual character of the history of technology. I wanted to learn whether these historians fostered the myth of progress or whether they were slowly creating a new language that could liberate us, in the West, from the passivity and violence of the myth. The expression "Whig history" began to attract my attention. It seemed to define precisely what I hoped to see overcome, a historical interpretation of technological change in the mode that Reinhard Rürup has called a "company history" where a body of technological success stories are told as if their final outcome were inevitable from the beginning. It was encouraged by what I found, although, as will be clear in the summary reflections of chapter 5, the evidence in TC is a mixed affair.

The procedure I have adopted is a detailed analysis of every article published in TC from its first issue in 1959 through the last issue of 1980. This approach has significant assets and limitations. By confining the study to articles in TC, I have omitted pertinent articles in European journals for the history of technology and in other historical journals in this country and abroad. I have also omitted an impressive group of monographs that has begun to appear in the past fifteen years, an omission remedied in part by a brief survey of monograph trends in chapter 5. Despite these limitations, however, TC articles provide a single body of historical research that is uniquely helpful for understanding the present state of the discipline. TC's editorial policy has been remarkably catholic, and the published articles reflect a broad range of methodological points of view. TC is recognized by many historians of technology, in the United States and elsewhere, as the field's major publication. Even more important, a significant number of TC authors are also active members of SHOT. They not only exchange new research at the society's annual meetings and in the pages of its journal, but they serve on the society's committees, review books, and referee manuscripts for TC. Because of their active involvement in the society, many of the authors know one another personally. Their network of relationships is a major contribution to shared scholarly discourse. It can be argued, therefore, that the complete corpus of TC articles is a representative body of historical research in the new field and that, as well as any limited sample can, it will reflect historiographic trends that have emerged in the field.

It is important to stress at the outset my priorities in designing this study. Because of the youth of the society, I will often speak of the birth and early development of the history of technology. Nevertheless, it is not my primary intention to study the formation of the discipline. While I have not ignored the background and personal relationships among the society's leading figures, neither have I exhaustively researched that area. My chief concern has been to study the language, the themes, and the methodological styles that set the intellectual tone for TC articles when they are seen as a single body of literature. I am more concerned about the intellectual character of the technological language found in the new discipline than I am about the sociological factors that have influenced its development. For this reason, chapter 1's analyses of the background of the society, and the more detailed treatment of these same matters in the dissertation on which this work is based (*Design and Ambience: Historians and Technology, 1958–1977*, University of Pennsylvania, 1980), should be understood as an attempt to provide the reader with a frame of reference for the textual analyses found in chapters 2 through 5.

It is equally important to note that I have not attempted a rigorous statistical study of the contents of TC articles. My taxonomic analyses of methodological styles, research parameters, and thematic hypotheses were not designed to be completely replicable by trained scorers. Indeed, a central premise on which this study rests is the conviction that the close reading of the text attempted here depends more on a slowly acquired intimacy with the text itself than on a quantitatively rigorous set of scoring categories. But I have not ignored all canons of textual analysis in creating the taxonomies that follow. I am indebted to three distinct intellectual traditions: cognitive anthropology, a historically based study of Western epistemological traditions, and hermeneutics. The nature of these three traditions is such that the reader should not expect statistically elegant computations of the taxonomic data. I reveal my debt to statistical theory only by my attempt to avoid exaggerated claims on the basis of an interpretation of the text that is primarily qualitative and personal. I suggest that the most helpful criteria for evaluating my conclusions will be the following questions. Do historians of technology find my detailed reading of the texts to be congruent with their own sense of the field? Does the reader find the major hypotheses proposed to be helpful and evocative for understanding the nature of technology in a historical context? Finally, do these hypotheses provide a helpful matrix for understanding the task of the historian of technology at present?

Several appendixes are included to assist the reader in evaluating the study's argumentation. The first discusses the design of the several kinds of taxonomies used in chapters 1 through 4. The second presents the central taxonomy of the study, an analysis of the methodological styles found in TC articles. Finally, appendix 3 is a single long taxonomy identifying each article judged to be part of the subordinate themes discussed in chapters 2 through 5. At every stage of the argument in these chapters I cite specific articles exemplifying each historical approach. It is important, when reading such evaluations, that the reader have access to the bases of my judgments. Thus, if I argue that one article is typical of many others, the reader can refer to the appropriate section of appendix 3. Each section in appendix 3 cites the pertinent pages in the text so that the appendix can be used as a thematic index.

The interpretation of the articles, beginning in chapter 1 and continuing through chapter 4, takes the following steps. Chapter 1 includes an analysis of the methodological styles found in the articles. It is followed by an essay that articulates a model of historical research as a communal rather than an individualistic endeavor. The chapter concludes with an exploraton of the patterns of research site selections—the specific time periods, geographical areas, and types of technology chosen as parameters for each article—to determine whether TC authors reveal any consensus about these three aspects of technology's historical terrain.

Chapters 2 through 4 explore the three most common thematic issues found in TC. Chapter 2 studies emerging technology, the single most popular theme in the journal and one that has generated the most explicit shared discourse. Chapter 3 addresses the pervasive but confused discussion of the science versus technology theme. A detailed analysis of pertinent articles leads to the surprising conclusion that the science–technology relationship is a false question for historians of technology. Its popularity is due to their concern to establish the irreducibly distinct nature of technological knowledge against a claim that modern technology is nothing more than an application of scientific knowledge.

Chapter 4 treats an even more confusing thematic issue, the relationship between technology and its cultural ambience, and concludes that TC's treatment of this relationship is remarkably underdeveloped at present. The chapter discusses three themes—the transfer of technology, technological determinism, and technological momentum—which appear in journal articles and show some promise of generating shared discourse about the technology-culture relationship.

Chapter 5 serves several purposes. It is meant as a summary of all major findings in the preceding detailed studies. More centrally, it explores the pivotal question of the entire study, whether SHOT historians have succeeded in escaping the myth of "Whig history," a myth based on the ideology of autonomous technological progress. Evidence from TC articles is presented to support both the argument that SHOT scholars foster a Whig interpretation and the argument that they have begun to create a new interpretation. A brief survey of some of the more important recent monographs, including all the books receiving SHOT's annual Dexter Prize, treats the same question. Finally, the chapter concludes by proposing an inclusive model for grounding all research in the field on a contextual integration of design and ambience rather than on the ideology of autonomous progress.

A word should be said about my choice of the expression "design-ambience" as a key term throughout the study. Since "ambience" might seem to be unnecessarily abstruse, a brief explanation is in order. Webster's *Third International Dictionary* defines ambience as "a surrounding or *pervading* atmosphere" (my italics) and ambient as an adjective meaning "surrounding on all sides." The expression is purposely elusive. It is also salient because of its relatively rare usage. It is important that the word call attention to itself and that it not be easily restricted to a narrow meaning. The definition of ambience as "pervading atmosphere" is particularly helpful because I use the expression as something more than a contiguous envelope around technical designs. For the purposes of the argument pursued here, a term is needed that includes two meanings: (1) a context that is not the technology in question, and (2) a context that permeates the design in question.

I refer repeatedly to the tension between technical design constraints and ambient factors, but there would be no tension in any meaningful sense if the two aspects were juxtaposed as container and contained. It is critical that their relationship be understood as "pervasive" and interpenetrating. Finally, the expression carries nontechnological overtones. Webster's *New Collegiate Dictionary* defines ambience as related to the decorative arts. By joining the expression "technical design constraints" with the decidedly nontechnical "ambience," it is my hope that the tension between design and ambience, so often referred to, will be understood to be as complex and elusive in historical analysis as it is in real life, a tension that reveals the many ways in which technological praxis is inextricably part of the larger human fabric.

It is difficult to study a body in motion, in this case the Society for the History of Technology, and to determine how it will develop in coming years. Thus, the concluding proposal at the end of chapter 5 must be seen as an educated guess about the discipline's methodological future. Given the importance of the cross-cultural starting point motivating this study, that same proposal should also be understood as my hope for our future direction. However, just as the present study of SHOT's intellectual character has been based on a study of existing research, so too the best interpretation of how historians of technology will proceed in the next decade will be not this educated guess but another retrospective study conducted some years from now. The host of new monographs in the field, together with new trends in published articles and papers at SHOT's annual meetings, is a strong argument for conducting such a study.

It is my hope that this analysis will prove helpful to my colleagues as a comprehensive interpretation of the field's intellectual journey from 1959 through 1980 and that it will provide enriched understanding of how we practice our art. Should my colleagues find this reading of their own work helpful, the study may also serve as an introduction for nonhistorians or beginning historians to the creative and vigorous discourse that has already begun to emerge, a discourse that provides critically important concepts and themes for interpreting the technological heritage of the West and for discussing the pressing issues of our common technological present.

Acknowledgments

Reading a book's acknowledgments is tedious if one is unfamiliar with the human richness embodied in what is essentially a list of names. For me, however, naming friends and colleagues who have supported the research and writing endeavor is a simple, essential vehicle for articulating a whole history of affection and gratitude. The temptation is to write another book to capture the variety and depth of those relationships that have deepened during the labors leading to the finished work. It is with a sense of the poverty of this format that I name my names and express my thanks.

It is impossible to estimate how much this study owes to the Oglala Lakota (Sioux) of Pine Ridge, South Dakota. Through their welcome over many years, the Lakota taught me to reverence the beauty of a non-Western people. Living on Pine Ridge also raised the question of Western Technology's impact on a small and delicate culture, the question that led me to the research presented here. My dissertation director, Thomas Parke Hughes, has followed my thinking processes from the outset with care, critique, and durable interest. Merritt Roe Smith has read the entire manuscript and suggested revisions with the care that only friendship can offer. Murray G. Murphey, my academic adviser during doctoral studies at the University of Pennsylvania, offered challenge and understanding at many critical points. Robert E. Kohler proved to be a thoughtful reader. Melvin Kranzberg, editor of *Technology and Culture*, allowed me full access to the journal files, read the first complete manuscript in critical and helpful fashion, and gave me a great deal of other help and encouragement. I. B. Holley and Father Robert Kearns, S.J., also read the manuscript and offered many helpful suggestions for revision. Leo Marx and Evelyn Fox Keller advised me in the final stages of the work. Arnold Thackray provided helpful bibliographical suggestions. Many of my colleagues in the history of technology have cooperated by participating in interviews or responding to questionnaires. I was also blessed with an exceptionally perceptive referee, W. Bernard Carlson, whose suggestions for revision have substantially improved the final version.

The Program for Science, Technology, and Society at Massachusetts Institute of Technology has been most generous in its support of my research. Together with colleagues like Roe Smith, Evelyn Keller, and Leo Marx, the support staff and graduate students proved to be a warm and stimulating ambience for research and writing. Colleen A. Dunlavy and other graduate students went out of their way to study and challenge my research. Lynn Roberson read the final manuscript with thoughtful critism and a careful editor's eye. Ann Serini was invaluable in helping to prepare the manuscript.

Marthenia Perrin and Rose Smith, administrative assistants for the departments of History and Sociology of Science and of American Civilization at the University of Pennsylvania, were frequently of great service as were Sylvia Dreyfuss, Patricia Johnson, and Elizabeth Cooper of the Department of History and Sociology of Science. Thomas Carroll spent many hours helping me design a computer program for organizing the data on which the textual study is based. Jeffrey Sturchio and George Danko both offered advice and occasional bibliographical suggestions. The warm and mutually helpful atmosphere of the Department of History and Sociology of Science, which permeates faculty, secretaries, and students, constitutes a marvelous environment for research.

The administration of the University of Detroit and, even more, my Jesuit community there understand and respect the demands of research and writing. They have consistently encouraged me and made sacrifices to help me.

Several close friends, Sister Consuela DeBiase, C.S.J., Sister Cynthia Comiskey, C.S.J., Sister Judith Kubish, C.S.J., and Sister Margaret Betz, O.L.V.M., assisted me during different stages of the work. They and many other Sisters of St. Joseph and Sisters of the Holy Child Jesus have repeatedly supported me with their friendship. A number of my Jesuit brothers, especially Christopher Mooney, Gerhard Böwering, Joseph Simmons, John Foley, Bruce Biever, Patrick Burns, George Casey, Robert Doran, James Ayers, Joseph Labaj, Art McGovern, Gerry Cavanaugh, Edmund Miller, and the men who have lived with me during the writing process, have shared this project at close range.

Finally, my parents, Louis William and Hildegarde Staudenmaier, and my immediate family have believed in my attempt at scholarship at least as long as I have. I am saddened that my father did not live to see the work completed, for his wisdom and encouragement are a dear inheritance. Without the love and cooperation of these people this book could not have come to be. I am very grateful.

Technology's Storytellers

1

The Society and Its Journal: The Emergence of Shared Discourse

We [John Rae, Carl Condit, Tom Hughes, and Mel Kranzberg] thought that an appropriate strategy would be to approach the History of Science Society and see if historians of science might widen their purview to include the history of technology. It so happened that Henry Guerlac, a leader in HSS, taught at Cornell. So a deputation went to see him. The meeting proved to be a disaster. . . .

We were crestfallen as we walked down the hill from Guerlac's home in Ithaca. "Well," I said, "if the History of Science Society is not going to 'condescend' to include the history of technology and if Isis *is not going to publish any articles dealing with it, then maybe we ought to form a society of our own which would concentrate on the history of technology and start our own journal." The others agreed, but, they said, "It's your idea, Mel, so you do the work."*—Melvin Kranzberg[1]

The year was 1957; the occasion the annual meeting of the American Society of Engineering Education at Cornell. Kranzberg, Rae, Condit, and Hughes, all members of the History of Science Society, met informally to consider ways to promote the history of technology as a new scholarly endeavor. Only after their unhappy encounter with Guerlac did the idea of an independent society take shape, and so, from a rib out of the side of the History of Science Society, SHOT (the Society for the History of Technology) was born.

Two challenges faced the leaders as they prepared, a year later, to publish the first issue of *Technology and Culture.* Would their society prove viable or would SHOT and the new journal linger for a time on the intellectual scene before succumbing to the harsh demands of institutional survival? More central still, would the question that drew them together—the historical interpretation of technology—retain its vitality and mature into a living body of shared discourse? Would that discourse win new participants to the little group? Both issues remained in doubt in SHOT's first years.

Around 1963, however, both society and journal began to show evidence of emerging from their early frailty into a durable identity. The story of SHOT's success in meeting both challenges, structural and intellectual, is the subject of the first chapter. They will not, however, be given equal weight. The central purpose of this book is to search out and articulate the language that historians of technology have created to interpret the technological past. The organizational history of the society will be treated in summary form while the lion's share of the chapter will explore the intellectual foundations of the discourse we find in the articles of TC's first two decades.[2]

The Society's Constituencies

SHOT's early development depended on four key constituencies, distinct in function, although blurred in their boundaries by overlapping membership. Two—the subscribing readership and the financial supporters of the venture—while necessary for survival, were less significant than the core group of society leaders and the larger constituency of contributing authors. TC's subscribing members numbered 688 in the first full year of publication. They grew at a healthy annual rate of slightly more than ten percent through 1967, when the rate of growth stabilized at its present level of approximately five percent. The contributions of the second constituency, SHOT's financial contributors, were vital during its first decade. Individuals and foundations offered grants to support society functions and to make up TC's operating deficits, which persisted until 1972, the journal's first year of modest profitability.[3]

In its early years the society's leadership constituency numbered fewer than twenty regulars. Almost all wrote articles for TC. They served on key SHOT committees, planned annual meetings, reviewed books, evaluated articles, and developed relationships with other scholarly and technical societies.[4] The tasks performed by the group were less important, perhaps, than the dynamism emerging from their enthusiastic commitment to the new field. As we shall see, the founders did not readily agree about the scope of journal offerings or, more important, about the definition of "the history of technology" itself. The vigor of their debates and the passion with which they pursued their sometimes conflicting goals provided an intellectual excitement that, more than any other factor, insured the survival and growth of the society.

From 1964 on a substantial number of younger scholars were attracted to the society. Successful recruitment of new leadership was a critical test of the society's long-term viability, and the process begun in the mid-sixties continues to the present. Typically, a potential leader begins contributing to the society by reading papers at annual meetings, by reviewing books for the journal, and eventually by publishing articles and serving on key committees. It is difficult to indicate exactly how many individuals should be considered leaders because no two people relate to the society in the same way. Nevertheless, the evidence suggests that this critical constituency more than tripled to approximately seventy-five members by 1980.[5]

Contrasting the early group who published articles before 1964 with those who have published in the last ten years reveals the changing nature of the author constituency. Evidence from journal correspondence indicates editor-in-chief Melvin Kranzberg had to work with a very small backlog of manuscripts until mid-1963.[6] Sixty authors wrote the sixty-eight articles published in the first four volumes. Nearly a third of the group, including most of the leaders of the society, identified themselves as historians of technology, of science, or of both.[7] Another fifth were historians from nontechnological areas. The remainder, nearly half the authors, represented a wide range of nonhistorical disciplines. Technical perspectives, such an engineering, science, economics, and business, tended to dominate, while the broad cultural perspective of the social sciences and humanities was less well represented.

The early bias toward engineering and science was even more pronounced than this evidence suggests. The idea of a society for technological history emerged at a meeting of the American Society for Engineering Education; many of the original members were active in the History of Science Society; and the society maintained active relationships with the History of Science Society, the American Association for the Advancement of Science, and numerous engineering societies.[8]

Several other characteristics of the early author group are noteworthy. Marjorie N. Boyer was the only woman to publish an article in the first four years, an indication that SHOT was an almost exclusively male organization at the outset. Scholars from outside the United States, particularly non-English-speaking authors, were a striking minority. Kranzberg recognized this lack and periodically made efforts to broaden the international scope of the author group, but with little success.[9] SHOT leaders wrote nearly one-third of the articles themselves, and several of the group, most notably Kranzberg, actively solicited contributions from other scholars. Kranzberg asked many well-known individuals for manuscripts and negotiated sets of articles published as single-theme issues in the fourth number of each volume.[10] Efforts like these led to an impressive array of articles—and representation on SHOT committees—by scholars who had become recognized in the second third of the century for pioneering studies of technology. They included Lynn White, Jr., William Fielding Ogburn, Cyril Stanley Smith, John B. Rae, Peter Drucker, Lewis Mumford, Howard Mumford Jones, Roger Burlingame, Abbott Payson Usher, S. Collum Gilfillan, A. Rupert Hall, A. Zvorikine, James Kip Finch, and Jacques Ellul. While most of this group were relatively inactive in SHOT affairs, White, Ogburn, Smith, and Drucker would serve as presidents of the society. In particular White, Smith, and Rae threw themselves into all aspects of the society's affairs.[11]

When we contrast this early group with a profile of the most recent ten years we see several significant shifts. The percentage of articles written by historians of technology or science rises from 32 percent to 48 percent. The overall percentage of historians increases from 59 percent to 72 percent. Indeed the influence of the historical approach on TC's authors is understated by this profile. As we shall see, the percentage of articles written in a nonhistorical format drops dramatically after the society's early years.

By the end of its second decade, the society and the journal had settled into a pattern of sustained membership growth and modest fiscal security, and they had moved more and more in the direction of a specifically historical community of scholarship. While there remained considerable openness to nonspecialist contributions, the growing percentage of articles written by members of the new profession suggests that the history of technology was beginning to dominate the journal's offerings. As it matured, TC more completely reflected the organization's name: the Society for the History of Technology.

This summary of organizational growth suggests that SHOT and TC have served as a forum for historical discourse about technology and that scholarly exchange within that forum has begun to generate a coherent intellectual focus for the field. Failure to find such a focus would have been fatal for SHOT. It would not serve to say, year after year, that the topic, the history of technology, was "interesting." Unless the members of the society began to interact so that the field took on an intellectual life of its own, the conclusion would correctly be drawn that the topic was sterile precisely because it had generated little shared discourse of moment. The continuance and steady growth of the journal and its relevant constituencies comprise evidence that this process of intellectual self-definition has in fact been taking place.

This conclusion is substantiated by one of the major findings in an author survey—a combination of personal interviews and written questionnaires—conducted in 1977. Ninety-five of the 191 authors who had published in TC responded. By far the most common response to any question in the survey was that indicating that TC has created a forum in which scholars of diverse backgrounds, often working in isolation, can publish and interact. One reply is typical of many. "T&C has been very important in giving scholars a focus for their research. This has helped to form a community of scholars."[12]

The Formation of a Viable Intellectual Focus

From the first, SHOT's struggle for intellectual identity has been rooted in a debate about method. Given the diversity of approaches to technological scholarship extant at its inception, it is not surprising that the society's members found consensus difficult. The question, as seen in SHOT's correspondence and in TC articles, was how to combine the two competences—technical and historical—whose interrelationship was claimed by the title of the journal.[13] Those who insisted that technical competence, as found in the best internalist research, was a sine qua non for history of technology argued that an excessively broad definition of the field would invite the publication of "soft" articles whose lack of technical competence would tarnish the reputation of the journal.

The demand for technical sophistication was more than matched by arguments against a pure internalist style. Several historiographical articles in the first three volumes pointed out that exclusive emphasis on design neglected complex interactions between the design and its context. All too often, it was argued, technological history was limited to success stories implying an autonomous technological determinism. The new society could meet a critical need by providing historical analysis of the failures, contingencies, and multifaceted components of technology's societal ambience.[14]

The two positions did not necessarily exclude one another. Advocates on both sides tended to criticize the other as a pure position, and it is true that some articles published in the first four years embodied pure internalist or pure externalist positions. Nevertheless, the arguments for the two approaches did not rule out an integration of the two after the manner of historians such as Lynn White, Jr., or Louis Hunter. In fact, although it was far from clear during these early discussions, the society's majority position, as seen in the body of published articles, would eventually favor the difficult integration of technical detail and its context.

The difficulty of this integration is illustrated by a striking parallel. In the first article of the inaugural issue of TC, Kranzberg explained the scarcity of extant literature studying "the development of technology and its relations with society and culture."

Serious historical scholars, with but few notable exceptions, shied away from the field because of a feeling that they lacked *the requisite technical knowledge* to treat it properly . . . [but] just as few historians are learned in technology few engineers are skilled in the *rigors of historical research.*[15] (my italics)

The contrast between "requisite technical knowledge" and "the rigors of historical research" is provocative because it is mirrored in technological praxis itself. It is clear that both the designing and the maintenance of technological artifacts demand detailed attention to functional design constraints. It is equally clear, however, that technological practitioners who design or maintain such artifacts are deeply influenced by the often ambivalent contextual factors that constitute the artifact's ambience. Technological activity does not occur outside this tension between design and ambience. Thus Kranzberg's tension between technical and historical expertise is directly parallel to the defining tension of technological praxis itself. In the words of Lewis Mumford:

History as the interpretation of the changes and transformations of a whole culture must necessarily take account of technology as one of the essential components of a culture, which in the very nature of the process affects, and is affected by, the pressures and the drags, the movements and resistances, the creativities and torpidities of every other aspect of society. By the same token, the historian of technology will find his account of technical processes seemingly isolated from the general flux of events, far more significant when he *restores technology itself to its dynamic social context.*[16] (my italics)

This discussion, helped as it is by two decades of development in the field, presents the tension between technical design and historical ambience in clearer terms than were available to SHOT members at the time. Their decision to name the journal *Technology and Culture* can be seen in hindsight as a decision favoring the integration of internalist analysis with contextual method.

TC's historiographical articles show that the debate continued well beyond the early years. Only twice in the first twenty years do we find a substantial number of these articles. The first cluster dominated the early issues of 1959 through 1961. The second appeared in 1974. In that year's first issue a set of three articles by R. P. Multhauf, E. F. Ferguson, and E. T. Layton, Jr., addressed major historiographical questions.[17] In the following issue European historian Reinhard Rürup published a parallel study of the field from a European perspective. Taken together, these articles provide a glimpse of the status of the field's intellectual self-definition in the early seventies. The articles written by Multhauf, Ferguson, and Rürup are of particular interest here because each attempts to assess the present state of the art.[18]

The central themes in all three articles are remarkably close to the core issues of SHOT's earlier debate. Multhauf and Ferguson stress the lack of a "conceptual framework" for the field. They acknowledge the existence of a body of specialized internalist history but point to the lack of unifying concepts to integrate the several technologies within a single universe of discourse.[19] Multhauf calls attention to the small number of internalist scholars responsible for a great proportion of the French, British, and American multivolume histories of technology that had appeared in the past two decades. It would be, he notes, falsely optimistic to infer from the existence of the three works that internalist scholarship was in anything more than its early stages.[20]

Ferguson and Rürup, on the other hand, stress the importance of developing a set of thematic questions that link internal design with societal context. They identify key issues, such as the nature of innovation, the validity of the concept of autonomous technology, the relationships between technology and science, between technology and economic forces, and between technology and other cultural forces.[21] In his penultimate paragraph Rürup summarizes the challenge facing historians of technology in terms remarkably close to the "design-ambient" tension.

> The social relevance of the history of technology lies—both for the present and for the future—in its critical function. We can use it to learn to distinguish between *technological necessities* and conscious and unconscious *social decisions*[22] (my italics)

In these three articles we find the two sides of the early debate—the need for technical competence and for historical integration of design and context—to be as central to historiographical discussion in the early seventies as they were a decade earlier.[23]

Further evidence of SHOT's methodological tension between design competence and historical sophistication can be seen in the author survey of 1977. The most common perception of the authors who discuss the "caliber of TC's scholarship" is that journal articles are of mixed quality even though TC is clearly the best journal in its field in the world. Many authors note "the broad range of topics and research styles in the articles" but their evaluations do not fit a single pattern. One group indicates that breadth of coverage is one of TC's best qualities, while the other considers it a distinct liability. Those who criticize the presence of too much internalist research are only slightly more numerous than those with the opposite complaint.[24]

An array of opinions about "the intellectual character of the field" recapitulates the main lines of the debate. Author perceptions on this issue fall into three clusters. The largest group refers to the difficulty and necessity of integrating what are seen as two distinct styles—technical design and contextual history. One author describes the greatest weakness of the field as "the gap between internal technical history (e.g., the lathe) and social-impact-of-technology history." Many of these respondents also indicate that TC's broad variety of research styles is the most important factor in the evolution of such an integration. Two smaller and equal-sized groups argue that the history of technology demands one of the two competences without reference to the other.

This survey of TC's authors taken near the end of two decades of the society's existence confirms once again the central hypothesis advanced here. The Society for the History of Technology began in 1958 as a small group of scholars with an extraordinary range of professional backgrounds. They lacked and felt the need for a forum within which they could meet one another and so begin a process of disciplinary shared discourse. While some held polar positions that can be called "internalist" and "externalist," their mutual participation in the single arena provided by SHOT created an intellectual climate fostering the difficult integration of the intricacies of technical design with the complexities of the cultural ambience of such designs.[25]

Scholarly Ancestry

In the thirty years before the birth of SHOT, research that would play an important role in shaping TC's methodology had begun to develop along three separate lines. For our purposes they can be termed "internalist history of technology," "nonhistorical analyses of technology," and "contextual history of technology." The internalist tradition is directly descended from centuries of European scholarship, but the second and third lines of research are much more recent and appear to be the result of initiatives beginning in the United States. Before we analyze all three in the context of the United States, however, it will be helpful to trace the development of internalist research in Europe and to consider briefly several other continental forms of scholarship, which have had less impact on TC's methodological formation.

"Internalist history" receives its name and heritage from the centuries-long tradition of interest, and indeed fascination, with the design characteristics of human mechanisms. It is "internal" history because the focus of attention is centered almost completely on the artifact itself rather than on how the artifact relates to its external social context. Its ancestry can be traced as far back as Giovanni Tortelli's *De orthographia dictionum e graecis tractarum* of 1449. Tortelli's work stands at the beginning of a long line of histories and encyclopedias of inventions culminating in von Poppe's *Geschichte aller Erfindungen und Entdeckungen* of 1837. Von Poppe categorized technological subject areas in a way that proved influential until well into the twentieth century.[26] After a hiatus of nearly fifty years we find evidence of renewed interest in the history of technology in the work of German scholars such as Amand Freiherr von Schweiger-Lerchenfeld and Ludwig Darmstaedter. Their individual contributions presaged a modest flowering of German interest marked by the beginning of work on the now-famous Deutsches Museum in 1903, the formation of two journals, *Beiträge zur Geschichte der Technik und Industrie: Jahrbuch des Vereins Deutscher Ingenieure* (Berlin, 1909) and *Archiv für die Geschichte der Naturwissenshaften und der Technik* (Leipzig, 1909), and the internationally recognized work of Conrad Matschoss, Ludwig Beck, and Franz Maria Feldhaus.[27]

The founding of the Newcomen Society for the Study of the History of Engineering and Technology in 1920 provided an institutional matrix for research in Britain. British research, unlike the German movement, survived during the World War II with the continuous publication of the *Transactions of the Newcomen Society*. After the war continental scholarship experienced a renaissance in France, Germany, Sweden, and Italy. Scholars of the first rank such as Bertrand Gille, Maurice Daumas, Torsten Althin, and Friedrich Klemm were of particular importance. Soviet history of technology has until recently had little connection with the traditions just mentioned. The work of the Institute of the

History of Technology at Leningrad from 1929 to 1937 was strongly influenced by its commitment to Soviet national pride and a Marxist interpretation of the relationship between technology and socioeconomic change, which sharply distinguished it from other scholarship. In the United States, as I shall indicate, internalist historians participated in the larger European universe of discourse. Thus the internalist methodology that would form part of the immediate ancestry of TC's intellecutal style was common to both sides of the Atlantic.

By contrast, nonhistorical analyses of technology based on sociology, economics, and political science have taken distinct paths in Europe and the United States. Neo-Marxist scholars, such as Jürgen Habermas and Herbert Marcuse, have had greater impact in Europe than in the United States. They form part of a larger European trend toward studies of the relationship between technology and society in terms of inclusive and elaborately articulated models, the most famous being Jacques Ellul's analysis of the deterministic power of "la technique." American studies of technological impact, such as Lazarfeld's analysis of the influence of television on children, tend to be more narrowly focused on particular technologies or analyses of the economic ramifications of innovation.[28] Finally, there appear to be few European parallels to what is described below as "contextual history of technology" which antedate TC's first volume.

The following analysis of TC's three immediate methodological ancestors is aimed at revealing the inchoate state of the art in 1959 and at underscoring the tensions within the early society in its search for intellectual identity. Each analysis is limited to a few important works, which are presented as exemplars of the essential characteristics of the three styles.

Internalist history

Between 1954 and 1962 major multivolume histories of technology were published in Britain, France, and the Soviet Union. These works represented the dominant style of the field at the time of SHOT's formation.[29] Internalist history was practiced by small clusters of scholars and antiquarians specializing in the design characteristics of single types of technology. Although interaction within such groups was vigorous, the several clusters tended to function in isolation from one another. "Technology" as a general term calling for collegiality across particular technological lines was not an operative concept in most internalist history.[30]

The History of Technology, edited by Charles Singer, E. J. Holmyard, and A. R. Hall, exemplifies this style. It defined technology as "how things are commonly done or made" and "what things are done and made."[31] The five volumes resemble an encyclopedia of technologies whose several historical developments are treated independently of one another. The "historical developments" of interest tend largely to be abstracted from the political and cultural fabric. Commentators on Singer's history and its parallels in France and the Soviet Union recognized that these volumes represented only a beginning of internalist work. The field was young, practicing scholars too few, and the works produced often suffered from a dearth of critical and collaborative research.[32]

Nonhistorical analysis

The work of Jacob Schmookler, William Fielding Ogburn, and S. Collum Gilfillan, although emerging from the two distinct fields of sociology and economics, represented a second style of research influential in SHOT's search for identity.[33] Each scholar, by choosing to make technology the centerpiece of his theoretical model, departed from the normal practice of his discipline. Most theoretical analysis in sociology and economics, when it dealt with technology at all, tended to relegate it to the status of a peripheral variable. By contrast, these scholars placed the patterns of technological change at the heart of their analyses. Gilfillan and Schmookler developed explanations of innovation.[34] Ogburn studied the relationship between new technology and societal values and structures. Their style stood in sharp contrast to internalist history. The demands of quantitative, systemic analysis common to both disciplines precluded attention to the design of individual technologies. For them, "technology" as a general socioeconomic force was more significant than individual "technologies."

Contextual history

Two books represent the third approach. Louis Hunter's *Steamboats on the Western Rivers* and Lynn White, Jr.'s *Medieval Technology and Social Change* created historical syntheses of technical design and historical context.[35] Both authors conceived of the internal design of specific technologies as dynamically interacting with a complex of economic, political, and cultural factors. Both historians—like Ogburn, Gilfillan, and Schmookler—stressed technology to an extent not typical in their fields. Historical analysis of the Middle Ages and of nineteenth-century America tended, in the main, to consider specific technologies as peripheral. For Hunter and White they were central. White's study created radically new interpretations of medieval social change in which a number of innovations related to mounted warfare, plowing, and power-driven machinery played the central role. Hunter's *Steamboats* broke from the typical internalist model by his extensive attention to political, economic, and social factors related to steamboat transportation.

Two earlier classics can be considered forerunners of the contextual approach. Lewis Mumford's *Technics and Civilization* (1934) articulated a complex, technologically centered interpretation of civilization in the West.[36] While his division of the stages of civilization into the eotechnic, paleotechnic, and neotechnic periods was based on technological factors, the study differed from White's and Hunter's by sacrificing some depth of detail in favor of exceptionally broad hypotheses. Even granting this limitation, the work remains an early representative of attempts to create a historical synthesis of technology and nontechnological factors. Abbott Payson Usher's *History of Mechanical Inventions* (1929) combined characteristics of all three styles. The body of the work traces mechanical inventions in a comprehensive summation of internalist historical research. In the first four chapters, however, Usher articulates a general theory of innovation not unlike Gilfillan's. Like contextual historians, he argues for the integration of technological development within a broad cultural context.[37]

Methodological Profile of the Journal

The debate about method is reflected in the variety of methodological styles—and in changing patterns among them—of the 272 articles published in TC between 1959 and 1980. To summarize our findings in advance, of the three contributing scholarly traditions to SHOT only internalist history and contextual history continue to play important roles throughout the first two decades of the journal. Nonhistorical analysis begins strong, but except for the influence of economic models on TC's interpretation of innovation, it dwindles to insignificance. This is, in fact, why the rather vague cover term has been adopted. "Nonhistorical analysis" does not differentiate among economic, anthropological, philosophical, or literary analyses because, although all are found in the set of articles, their several disciplines are less important than the fact that they are not historical scholarship. It appears, in other words, as if SHOT's gradual maturation as a historical society rendered these nonhistorical contributions less and less appropriate for TC's pages.

To interpret the complexities of TC's methodologies a taxonomy was devised scoring all articles in two dimensions: general style and use of hypotheses.[38] The following survey of these taxonomic findings

Table 1
Methodological styles in *Technology and Culture* articles, 1959–1980

General Style	
Contextual	136 (50%)
Internalist	47 (17%)
Externalist	37 (14%)
Nonhistorical analysis	32 (12%)
Historiographical reflection	20 (7%)
Function of Hypotheses in Argumentation	
A priori	151 (56%)
A posteriori	121 (44%)

n = 272.

will help to understand the methodological substructure of each article, will provide an overall profile of TC's changing methodological styles, and, most important, will create a matrix within which these remarkably diverse articles can be "read" for their thematic content. Without help from such a differentiating taxonomy, it is virtually impossible to interpret thematic patterns. Before we analyze the articles according to these methodological categories, it will be helpful to see them in a single overview (table 1).

Three of the five methodological styles—internalist history, contextual history, and nonhistorical analysis—have already been identified as the three dominant types of technological research existing before 1959. A fourth style has been named "externalist" because it is the exact opposite of internalist research. Externalist articles study the context of technological events but do not discuss the design of function of the technologies in question. A fifth group of articles does not represent a style of research so much as reflection on the nature of such research itself. For this reason these "historiographical reflections" should be thought of as a metastyle. They are best understood as an extension of the vigorous private discussion of the field's intellectual identity already discussed.

Eleven early historiographical articles all appear in the first seven issues, dominating journal offerings in a fashion never again repeated. Even the fact that they make up half of the first twenty-one articles does not fully indicate the level of TC's early preoccupation with such reflection. The entire Fall 1960 issue was devoted to reviewing the five-volume *History of Technology* edited by Singer, Holmyard, and Hall together with other recently published histories of technology from France, Italy, and the Soviet Union. It is possible, of course, that historiographic interest was not the only motive behind this remarkable concentration. It would be easier for scholars to write a short essay about their perception of the new field than to provide a full-scale article on relatively short notice. In fact, the articles are unusually brief, averaging ten pages compared with an average of over thirteen pages for all articles published before 1964. After this burst of publication the form disappears until 1970. Nine historiographical essays were published in the seventies, four in the cluster of 1974 and the other five separately. As noted, the appearance of the post-1970 articles, together with even more recent debate, indicates an ongoing interest of SHOT members in questions about the field's intellectual character.[39]

To see patterns of change among the other four styles it is helpful to break the twenty-one years of this study into three seven-year periods. The styles reveal, in their changing frequencies, some significant patterns of the role of hypotheses within each type. A scholar interacts with his or her peers most explicitly in two ways: by generating new hypotheses to interpret historical evidence and by critiquing or modifying existing hypotheses. In our taxonomy the category titled "*A Posteriori*" refers to articles whose authors are primarily interested in establishing one or more new hypotheses. The category named "*A Priori*" refers to articles whose authors explicitly respond to already-articulated hypotheses.

This perspective reveals the unique status of the nonhistorical analyses in TC. Only one of the thirty-two such articles generates new hypotheses on the basis of evidence reported in the article itself.[40] Nonhistorical essays tend to be summary statements of theories whose origins lie in other disciplinary communities. They can be seen to be the work of outsiders, not only because the vast majority of their authors were not SHOT regulars, but also because of the structure of argument in the articles themselves. It is clear that these articles are not addressed to SHOT members as a body of critical scholarly peers.[41] The marginal status of the style is further indicated by its virtual disappearance after the first seven years of TC's existence. Six articles appear in the middle seven-year period, and the number dwindles to two in the most recent period.

In constrast, the two largest clusters of articles—internalist and contextual—play an important role in TC's methodological development. In the first seven years they are nearly opposite in their style of argument. Seventy-one percent of the contextual studies generate new hypotheses, but they are almost always limited in scope to conclusions about a single event.[42] Thus Walter Dornberger's study of German V-2 rockets articulates hypotheses about the structure of the V-2 team and about relations with other German wartime institutions, but he does not attempt to expand these hypotheses beyond the V-2 case.[43] This cautious limitation of hypotheses is a sign of the youth of the field. These articles tend to be pioneering ventures into areas explored by few others. The questions being probed are by and large new questions. While the early contextual articles do not engage in much explicit dialogue, they serve the important function of identifying important thematic questions for the field. They do so, however, in a cumulative process by which highly specific hypotheses about individual cases begin to form a body of literature, which becomes more capable of generating explicit discourse as it becomes more extensive.

Early exceptions to the contextual *a posteriori* pattern demonstrate what is meant here by "explicit discourse." Robert Woodbury's attack on the myth of Eli Whitney as father of interchangeable parts manufacture, Milton Kerker's critique of the supposed independence of steam engine innovation from science, and Lynn White's assessment of Eilmer of Malmesbury's experiment with heavier-than-air flight all critique hypotheses that were articulated by earlier historians.[44] In so doing, they reveal the existence of clusters of scholars actively engaged in debate. It is this sort of mutually critical research that tends to be lacking in most early contextual articles. The number of historians adopting the contextual approach at this time was so small relative to the vast historical terrain open to them that most labored in research areas with no near neighbors. The historical process, described by Lynn White in his article on the nature of invention, could occur only in situations where the historian had peer critique.

> Since man is a hypothesizing animal, there is no point in calling for a moratorium on speculation in this area of thought [i.e., history of technology] until more firm facts can be accumulated. Indeed, such a moratorium—even if it were possible—would slow down the growth of factual knowledge because *hypothesis normally provokes counter-hypotheses*, and then all factions adduce facts in evidence, often new facts. The best that we can do at present is to work hard to find the facts and then to *think cautiously about the facts which have been found*.[45] (my italics)

White's invitation to do "the best that we can do at present," to look for facts, and to "think cautiously" about them, appears to be an accurate description of contextual history in the early years of the journal.

The role of *a priori* hypotheses in internalist articles is a significant indication of the advanced state of internalist as compared with contextual discourse. Fourteen of the twenty internalist articles in the early years respond to some explicitly stated prior hypothesis. Thus Marjorie N. Boyer attacks the assertion that the Roman pivoted axle had been forgotten during the Middle Ages.[46] Like the articles by Woodbury, Kerker, and White, these fourteen reveal the existence of highly focused historical discourse. The fact that so many early internalist articles take this form reminds us of the relatively large number of semi-independent pockets of internalist scholarship in existence by 1959. TC's publication of some of their research served a dual function. On the one hand, the journal became a forum in which many previously independent groups of internalists could interact with one another.[47] On the other, the journal's policy of publishing research in different methodological styles opened the internalists to interaction with other, radically different, viewpoints about appropriate methods in the field.[48]

While the frequency of and the preference for *a priori* hypotheses remained roughly constant for internalist articles throughout the second and third seven-year periods, several shifts in contextual scholarship are noteworthy.[49] Contextual articles increased their share of the overall articles, from 41 percent in the first seven years to 53 percent in the middle period. During these years the pattern of contextual preference for cautious *a posteriori* hypotheses remained unchanged.

In the seven years after 1973, however, two remarkable changes occur which indicate a maturation of TC's universe of discourse as a body of contextual history. On the one hand, contextual articles become the clearly dominant style (68 percent). Even more significant is an almost complete reversal of the proportions of *a priori* and *a posteriori* hypotheses. After 1973 the share of contextual articles adopting an *a priori* use of hypotheses jumps from 29 percent to 57 percent. This shift, revealing as it does pockets of contextual historians engaged in explicit discourse, merits further attention. What were the questions that had begun to generate such dialogue? Three themes stand out: ten contextual articles analyze cases of technology transfer,[50] eleven treat cases of innovation,[51] and seven assess theories of the science–technology relationship.[52] The remaining ten treat independent issues.[53]

The significance of this change should not be underestimated. It is an unmistakable sign that the contextual approach to the history of technology has emerged from its years of infancy marked by individualistic interpretations of single cases and has entered a more mature stage of development. The cumulative effect of the *a posteriori* hypotheses articulated in TC has been to focus attention on several themes that transcend the case studies from which they originated. In a recent historiographical article, Otto Mayr describes the communal process of historical research in precisely this fashion. "The historian's approach is fundamentally inductive rather than deductive; it *begins* with microscopic research in hopes that the empirical data thus gathered *will lead to generalizations on some higher level.*" (my italics)[54] The fact that well over half of the recent contextual articles engage in explicit shared discourse is a clear indication that the shift from microscopic research to higher-level generalizations has begun to occur. By their own research efforts, therefore, and with the help of the forum provided by SHOT, contextual historians have begun to define their intellectual identity in thematic terms.

The fourth methodological style found in TC, externalist history, provides added evidence of the dominant role of contextual history. Externalist research was slow to find its place in the overall array of articles. The ten articles scored as externalist in the first seven-year period tend to treat topics untypical of most TC research or to summarize prior research after the manner of nonhistorical essays.[55] In later years several patterns emerged. Externalist articles climbed from 9 percent to 18 percent of TC articles by the second period, and they held nearly that proportion, 16 percent, in the final period. In both periods their numbers exactly matched the number of internalist articles. The majority of the twenty-seven articles published after 1966 generated new hypotheses based on case studies. These *a posteriori* hypotheses are strikingly similar to the early contextual articles in that they tend to be limited to the case study in question. Thus Daniel Kevles's discussion of the post-World War II struggle for control of federally funded research offers hypotheses explaining why the military eventually prevailed. He does not, however, generalize beyond this case.[56]

These articles from the most recent fourteen years represent a new style of research in which historians study the ambience of technology without analyzing the design characteristics of the technologies in question. The emergence of this style is another striking example of the governing role of TC's dominant style, the contextual history of technology. The gradual accumulation of contextual studies has identified "the technological ambience," that is, a historical ambience precisely as it is related to technology, as a historical subject matter worthy of research in its own right. It is not surprising that the externalist style of history, taking the technological ambience as its methodological center of interest, should have developed some years after contextual studies. Before such research could become an accepted part of TC's shared universe of discourse it was necessary for SHOT members to recognize its importance. The recognition appears to be one result of the growing maturity of the contextual tradition.

These chronological relationships among TC's three historical styles are another sign that SHOT began in a predominantly internalist climate. Contextual articles only slowly break from an internalist fascination with technical design abstracted from social context and only slowly generate thematic questions as foci for explicit disciplinary interaction. Externalist articles, a radical break from the internalist tradition, develop more slowly still and their development appears to be mediated by the increasing maturity of contextual research.

The numerical balance in recent years between internalist and externalist styles suggests that both have taken the role of valuable adjuncts to the central body of contextual history being published in TC. In their pure forms both styles contribute new insights. Given the newly dominant definition of the history of technology as that discipline which attempts an integration of technical design and its ambience, both remain dependent upon the vigor of ongoing contextual research for the creation of a single shared universe of discourse in which they can participate.

The Communal Character of Historical Research: An Interpretative Model

By this point it has become commonplace to refer to SHOT and TC as a universe of discourse. Before treating the contents of TC articles in subsequent chapters, it will be helpful to explain the expression by considering the communal nature of history in greater detail. The following model has emerged during the long process of analyzing the articles and designing taxonomies to reflect article usage. It reveals my own intellectual heritage in its use of insights from three traditions that have influenced my thinking; cognitive anthropology, general epistemological theory, and the special branch of epistemology known as hermeneutics.[57]

Five modes of historical scholarship

It is helpful to think of historical research as a process in which the individual scholar interacts with existing perceptions of the past. In the process five distinct types of interaction can be identified—the choice of a site for research, the determination of priorities among types of evidence, the use of themes to interpret evidence, the assessment of validity of evidence and of inferences from it, and the process of explicit dialogue with other members of the historical community. We will discuss the five interactions sequentially, but care should be taken not to infer that they occur in simple linear progression. Historical research is a dynamic process punctuated by moments of choice when one or another of the activities comes to central focus.

The first aspect of research is the choice of what we shall call the "site." By its very nature historical research takes place within a temporal and geographical context, which must be specified at least minimally. Historical events do not exist as atemporal or unearthly abstractions. Frequently the choice of specific boundaries for the research site is not the first decision made in a new research project. Boundaries will be shaped and reshaped according to the demands of the evidence uncovered during the entire process. Nevertheless, at some point the historian becomes decisive about the limits of time and place that will enclose the specific interpretation being advanced. Setting boundaries for a research site is, therefore, directly related to decisions about what evidence is central to the study and which hypotheses best interpret that evidence. To take a simple example, let us suppose that a historian of nineteenth-century American politics has uncovered evidence indicating that the political and economic dynamic usually called the post-Civil War reconstruction was less dependent on the war than on forces that began to be influential as early as 1830. The decision to stress the new evidence as a basis for a new look at "Reconstruction" will be accompanied by a decision to redefine the temporal boundary of the period.

The determination of priorities among types of evidence is a second choice facing the historian. The array of potential evidence is almost always greater than the amount practical for a given purpose. This imbalance forces decisions of priority. Some evidence will be treated as central, some as peripheral, and some as irrelevant. Thus, if a historian were to focus attention on evidence pertaining to James Watt's struggles with patent law, s/he would tell a different story than if central focus were given to the influence of scientific theory on Watt's inventive insights.[58] Conceivably, several historians with different evidentiary priorities could use identical bodies of evidence and still write radically different interpretations. Of course such prioritizing decisions will be affected by choices of site boundaries and of particular thematic questions of interest.

It can be argued that the third aspect of historical research, the thematic questions that the historian chooses to ask of the evidence, is the most important creative act of the entire historical process. Lacking thematic interpretation, history is reduced to a laundry list of "facts." Indeed, the historian's primary contribution to an understanding of the past depends on the questions s/he decides to ask. It can also be argued, of course, that the thematic questions that interest each scholar will influence decisions about site boundaries and and the relative priority of evidence. Thus, in the example used above, the decision to focus attention on evidence pertaining to Watt's legal struggle is rooted in a thematic question about the role of patent law in the innovation process. The fact that thematic questions are central, however, does not imply that they necessarily precede other decisions. It is quite possible that an encounter with surprising new material happens before the historian decides what questions to ask of it. Thematic questions are central because they provide the intellectual rationale that gives coherence to decisions about site boundaries and types of evidence. We can see, therefore, that the first three aspects of the historical process are intimately related acts of choice. Taken together, they are the core of the individual historian's creative contribution to the larger culture's understanding of the past.

The fourth aspect, the assessment of validity, is of less concern to us here even though it is an essential prerequisite for research. The historian cannot escape the responsibility of verifying, as thoroughly as possible, the validity of each piece of evidence used to advance an argument and the validity of any inferences drawn from that evidence.[59]

The fifth aspect of research, the dialogue between the individual historian and the historical community, occurs when new research is disseminated, whether in print, at formal meetings, or in informal conversation. The critical response of other historians, whose expertise enables them to assess the validity and relevance of new research, is an essential component of the process.

Were we to conclude this description of the historical process here we would imply that the first four aspects are primarily individualistic, that the significance of the historical community appears only in the dissemination process. To so imply would be a crippling oversimplification. On the one hand, it would suggest that "the historical community" is limited to professional historians and, on the other, that these historians form a community only on the level of explicit discourse. In fact, the historical community is much more extensive than the historians themselves, and its influence pervades every level of the research process in both conscious and unconscious ways.

Historical scholarship within a cognitive world view

To underscore the full reality of what we are calling the historical community, we must conceive of it as the sum of all persons, professionals or not, who in any way influence a culture's perception of its past. More important, we must recognize that the very existence of this community is predicated on a cultural body of knowledge that includes and transcends the individual knowledge of its members. To explain the principle at work here it is necessary to introduce the anthropological theory on which it is based. Each individual decision by which the historian shapes new research takes place within the horizon of his or her previous understanding of the historical past. This previous understanding, received in the historian's earlier cultural education, functions as a set of assumptions about the past that will influence all subsequent decisions made during research. Finally, in the creative process of new research, the historian not only modifies his or her own previous understanding of the past but s/he also exerts some influence on the cultural community's understanding as well.

A central thesis of cognitive anthropology asserts that all individual activity occurs within a cultural ambience, a cognitive universe giving shape and meaning to the world. Multifaceted and complex, this cognitive universe is never appropriated by an individual in its entirety, nor does it function exclusively on the conscious level. James Spradley describes its role.

> Most people do not stop to consider that they are continually using categories as they think and talk. Much of their knowledge about the classification of experience and the attributes that are used for this purpose are *outside of awareness. They believe that it is natural for the world to be divided up and structured in the way they have learned it to be.*[60] (my italics)

If a person did not share in some culturally learned universe of discourse that "divided up and structured" the world in some meaningful fashion, it would be impossible to think or to act. Such is the basic premise of cognitive anthropology.

The fact that no individual completely comprehends the whole of the culture's cognitive universe is of particular importance.[61] One's grasp of the cultural world view is incomplete because culture is learned in a finite manner beginning from a unique and limited starting point. In other words, each person's cultural perception is radically conditioned by personal history. It can be argued that the common cognitive universe in which individual members share exists as the "linguistic basis" for discourse. In the broad meaning used here, the "language" of a culture includes not only the normal linguistic dimensions of vocabulary, syntax, and usage, but also a whole set of norms and values, of beliefs and rules for behavior, which the individual must understand to be considered "fluent" in the culture.[62] Because the cognitive universe of a culture is so centered in the human experience of communication, it is now helpful to expand our term "cultural ambience" to the culture's "shared universe of discourse." This cultural world view, which is first received passively, becomes the basis for all communication and indeed for all cognition. By participating in such a shared universe of discourse—both as recipient learner and as creative contributor—the individual becomes a full member of the culture's life process, a process in which the shared universe of discourse is continually modified through the creativity of its members.

This anthropological theory sheds considerable light on the interplay between the creativity of the individual historian and the larger historical ambience in which research occurs. The communal dimension of research is not based on the historical community seen as a collection of individual historians, as much as it is in the shared historical universe of discourse that serves as the basis for dialogue among the historians. In every stage of the research process the historian is in dialogue with that shared universe in precisely the same way that every participant in a culture lives in dialogue with his or her cultural universe of discourse.

Consider first how the historian is preconditioned by the common perception of history before ever beginning a research career. All of the history that s/he has learned—whether the formal research of professionals or the whole body of informal assumptions about the past and present which have been absorbed from one's family, the popular media, and so forth—operates as a set of biases preshaping the starting point for research. Like all other cultural presuppositions, these prejudices often exert influence without one's conscious awareness of their existence. They are the "natural" ways for the past to be divided up and structured. These unconscious prejudices, together with the explicit positions the scholar has come to hold about history, constitute the basis for all further research. This starting point is, in fact, a set of assumptions about the "orthodox" boundaries of time and place on the historical terrain, about the relative importance of various types of evidence, and about thematic questions that are worth asking.[63]

To speak of the "prejudices" of the historian is in no sense meant to be pejorative. To assume that the historian could approach the task of history unencumbered by prejudice (e.g., class, sex, national origin, race, culture, etc.) would be to assume that s/he was somehow outside of the historical process, having no personal history, no limiting starting point. Since this is impossible, such an assertion of objective and bias-free historical research would achieve nothing except to blind the scholar to every sign of personal and cultural presuppositions. Hans-Georg Gadamer states the principle aptly.

> A person who imagines that he is free of prejudice, basing his knowledge on the objectivity of his procedures and *denying that he is himself influenced by historical circumstance*, experiences the power of the prejudices that *unconsciously dominate him* as a *vis a tergo*. A person who does not accept that he is dominated by prejudices will fail to see what is shown by their light. . . . Historical consciousness in seeking to understand tradition must not rely on the critical method with which it approaches its sources, as if this preserved it from mixing in its own judgments and prejudices. It must, in fact, take account of its own historicality.[64] (my italics)

Given the importance of learned prejudices as the origin for new research, it is clear that any contribution will take the form of a modification of the existing universe of discourse. In the first instance, the historian modifies his or her own prior sense of history by choices made during research. In the second instance, s/he exerts some influence on the communal universe of discourse when publishing new findings.

Consider how this process might affect each of the three areas where historians must make decisions. It is possible that published research will question established geographical or temporal boundaries simply by articulating new ones. Insofar as the new definition of site boundaries is accepted by the historical community, the scholar will have modified existing assumptions. Research need not challenge existing assumptions. When new research accepts previous consensus about time or place boundaries, it strengthens the position of the consensus. The parallel challenge or confirmation of communal consensus about valid types of evidence is discussed by Howard Mumford Jones in an early TC article.

> He who is trained to *scholarly orthodoxy* tends to look for what he wants in *the right places only*; and the right places have been in records and documents. . . . A peace treaty, a poem, a painting, a system of philosophy, an anthropological report possess academic respectability; a lever or an ink eraser do not. Historical evidence is, and has been, curiously "literary."[65] (my italics)

Finally, it is clear that the use of thematic questions already articulated by earlier scholars will reinforce their influence in the field, and that the creation of new themes will act as a critique on the existing set of interpretative hypotheses.

A new field of history, such as the emergence of a formal society for the history of technology, affects this process in two ways. In the first place, the preexisting set of assumptions about the meaning and shape of technology's historical past is more inchoate than is the case for older branches of history. As a result, the historian of technology has more freedom in shaping the boundaries of research sites, in determining priorities of evidence, and in choosing themes. S/he is more free precisely because there is much less historical discourse to be learned. On the other hand, the support provided by a well-articulated branch of history is to the same extent diminished. It is more difficult to construct a comprehensive and insightful historical interpretation without the help of an existing tradition and without the challenge of vigorous peer critique from many established scholars who are "near neighbors" to the area chosen for study. In the second place, the creation of a forum for scholarly exchange, such as TC and SHOT, will intensify this entire process. The process of publication—in TC and at SHOT's annual meetings in particular—becomes a major force in generating a shared universe of discourse that will continually influence, and be influenced by, new scholarship.

This communal model helps to clarify the assumptions and goals of this study of TC articles. Evidence presented earlier in the chapter strongly suggests that TC and SHOT have begun to create a community of historical discourse about technological change. The evidence suggests in turn that the interactions—between individual historians and their emerging universe of discourse—characteristic of such a community have been occurring in the twenty-one years covered in this study. It is not unreasonable, therefore, to consider the 272 TC articles as a single "text" in which we can discover an unfolding technological language, a language that reveals the intellectual character and world view of the new historical community and at the same time introduces us to the beginnings of a new way to interpret technological change.

A thorough sociological analysis of SHOT's attempts at disciplinary formation is not the primary focus of this study, although attention to the relationships, in the act of writing history, between personal and cultural assumptions and the content of research is important to understand the inner workings of the society and the formation of journal policy. Nevertheless it is the precise character of SHOT's historical *language* that remains the central concern.

This is a difficult matter. It can be argued that a detailed sociological analysis of SHOT leaders—of their links with existing traditions in the History of Science Society, the American Association for the Advancement of Science, and a variety of engineering traditions, of SHOT's sources of funding, etc.—would reveal a conservative "world view" that bound the group to an inherently conservative "internalist" approach.[66] From this point of view contextualism, insofar as it demands a radical integration of Western technical designs with Western political and economic biases, would tend to be beyond the reach of SHOT's conservative perspective. In such a view, SHOT historians would be little more than Whig historians who chronicle the success story of Western technological achievements.

However, as will become clear in chapter 5, my study of TC's corpus of articles leads to a more complex interpretation. I argue there that TC authors have made impressive beginnings in the slow and difficult task of creating a contextual approach that has the potential to liberate the historical interpretation of technology from Whig history. Nevertheless, the detailed textual analyses of the next three chapters leads me to a critique of TC's blind spots which is not far from the more sociological analysis suggested above.

My intention in stressing textual, rather than sociological analysis is not to avoid the question of SHOT's biases, Whig or otherwise, but to permit the texts of TC's articles to speak for themselves. It is, finally, the *language* of technology's storytellers that exerts influence on the pervasive popular rhetoric of autonomous technological progress, either by legitimating it or by providing a liberating alternative. As I noted in the introduction, it is the cultural influence of "progress talk" that most concerns me and that led to this study. My conviction is that a fair and careful reading of the texts before us is the best approach to what is, at its core, a question of language.

The communal model provides several guidelines for textual analysis. As noted, the formation of new historical language takes place in the three kinds of creative decisions made during research: choices of site boundaries, decisions about priorities among types of evidence, and adoption of interpretative themes. The three will not be treated equally. We have already completed the primary analysis of evidentiary priorities by constructing the taxonomy sorting TC's historical articles into three types: those focused on the data of technical design alone ("internalist history"), those focused on contextual evidence alone ("externalist history"), and those attempting to integrate both types of evidence ("contextual history").[67] Subsequent chapters frequently use the matrix provided by this methodological taxonomy to shed light on thematic discourse in the journal.

Thematic interpretations receive by far the greatest attention in this study. The textual analyses of chapters 2 through 4 are designed to call attention to emerging consensus in TC, both expressed and implied.

Particular attention will be paid to what have been called *a priori* hypotheses because they reveal clusters of historians already engaged in explicit discussion of such issues. On the other hand, the relative youth of the SHOT community demands that we seek out patterns of usage which, while not yet developed into explicit discourse, have the cumulative effect of creating the basis for later such dialogue.

Care must be taken not to misunderstand the significance of the claim for thematic consensus being advanced here. The use of the expressions "consensus" or "emerging consensus" does not imply facile agreement among scholars on the march toward unanimity in their interpretations of technological change. "Consensus" here refers to the minimum of commonly shared language and assumptions that must serve as the basis for human discourse of any kind. As will be shown, the areas of linguistic agreement in TC sometimes suggest a high degree of ideological consensus, as in the consistent thematic treatment of the question of development, and sometimes the exact opposite, as in the sometimes contradictory interpretations of technology transfer.[68] Therefore the following reading of the "text" of TC articles serves three purposes: first, to provide evidence that historical discourse is in fact occurring in the forum provided by TC; second, to identify those thematic questions that have attracted the attention of significant numbers of the scholars who publish in TC; and third, to present a coherent overview of the interpretative language about technological change—with its ideological consonance and dissonance—that has begun to develop in the pages of TC.

Before turning to the thematic analysis in chapters 2 through 4 one task remains. As we have noted in our communal model, historians share in a discourse when they make choices of site boundaries for their research. What can we learn about TC's interpretation of technological history from their decisions in this aspect of the historical endeavor?

Land, Time, and Technology: Dimensions of Historical Terrain

The term "terrain" suggests the relationship of earth with maps. Whether or not the boundary lines of a map are based on obvious physical characteristics of the land, they necessarily represent some human consensus about boundaries. In like fashion, locating research in a specific time and place presupposes some set of temporal and geographical norms, presuppositions—either conscious or unconscious—about which places and times are historically more significant. Cumulatively, therefore, time and place choices within a community of historians establish normative boundaries giving a particular character to the terrain of the past.

While geographical and temporal dimensions are essential for every branch of history, a given historical specialty may also designate topical boundaries. For internalist history of technology topical categories, based on traditional divisions of technology types, have often operated in this fashion. The value of topical taxonomies has been seriously questioned by some historians of technology, but their influence remains important and can be seen in the major divisions of bibliographies in the field.

TC's articles have been scored according to these three dimensions, and the evidence revealed by the resulting taxonomies indicates several significant patterns. On the one hand, they allow us to see the extent to which TC's shared universe of discourse is based on site convergence. On the other hand, the same areas of concentration reveal presuppositions about technology's historical terrain that suggest implicit biases about the definition of technology itself. We will consider the three taxonomies individually and then consider correlations between time periods and geographical areas. In conclusion, we will note several characteristics of the three historical styles—contextual, internalist, and externalist—that appear when they are correlated with the time-period taxonomy.

The geographical pattern seen in table 2 is significant and unambiguous. With very few exceptions, the articles locate their research in "the West" as it is commonly defined by the term "Western civilization." The United States and Western Europe account for 80 percent of all place references. This Western bias is further enhanced by two smaller clusters referring to the Middle East and the Mediterranean basin. Both areas are perceived to be the direct cultural and technological forebears of Western Europe. By contrast, references to Asia, Latin America, Australia, and Africa comprise altogether under 8 percent of the references.[69]

Given the Western bias in geography, it is not surprising that time periods tend to fit Western categorical norms as well. The beginning and end dates in the seven periods seen in table 3 have been set to approximate usage in the greatest number of articles. Thirty-eight articles cover several periods. Many of these attempt overviews of Western technology from prehistoric times to the present. Others trace the history of one type of technology through a sequence of dated events. The relatively small percentage of multiperiod articles confirms the validity of the seven time periods selected because the remaining articles can be scored within their boundaries.

Table 2
References to place in *Technology and Culture* articles

	Core references	All references
United States	114 (47%)	133 (35%)
Europe	32 (13%)	53 (14%)
Britain	29 (12%)	54 (14%)
France	11 (5%)	25 (6%)
Germany	8 (3%)	13 (4%)
Russia	7 (3%)	10 (3%)
Other European	12 (5%)	13 (3%)
Greece and Rome	9 (4%)	26 (7%)
Middle East	6 (2%)	22 (6%)
Asia	8 (3%)	23 (6%)
Latin America	5 (2%)	6 (1%)
Africa	1 (1%)	2 (0.5%)
Australia	0	1 (0.1%)
Totals	242 (100%)	381 (100%)

Table 3
Time periods of *Technology and Culture* articles

Ancient (5000 B.C. to 600 B.C.)	6 (2%)
Classical (600 B.C. to 400 A.D.)	10 (4%)
Medieval-Renaissance (400 A.D. to 1600 A.D.)	26 (10%)
Scientific and Industrial Revolutions (1600 A.D. to 1800 A.D.)	21 (8%)
Nineteenth century	70 (26%)
Twentieth century	75 (28%)
Several periods	38 (14%)
No time references	26 (10%)

n = 272.

The pattern shown by the articles falling into a single period is clear: the more remote the period, the fewer articles in it. Several aspects of these frequencies are noteworthy. The Middle Ages and the Renaissance have been collapsed into a single period because of the small number of references to the Renaissance. Although eighteen articles are devoted to medieval technology, the six centuries between the decline of the Roman Empire in the West and the beginning of the second millennium A.D. are virtually ignored. It is also obvious that the nineteenth and twentieth centuries are the overwhelming favorites.

Topical references to types of technology are diffuse. This is reflected in the remarkably asymmetrical nature of the subcategories in table 4. Some refer to one specific technology (the italicized subcategories), others to clusters of related types of technology. We find a handful of small clusters of articles on topics such as steam engines, machine tools, rockets, and the internal combustion engine. None has been the focus of shared discourse in the full sense of the term—explicit critical interaction among scholars about the same technology in the same place and time period.[70] The potential types of technology that could be researched by historians are so many that this scattered and unfocused pattern is not surprising. Because the absence of pattern does not permit us to draw any conclusions about the presuppositions of journal authors, the following analysis will be limited to geographical and temporal references.

The correlation of time and place patterns in table 5 reveals several further aspects of TC's site selection profile. The primary geographical referent for thirty-three articles is "Europe," understood as a cultural region of the world transcending national boundaries. Such references dominate the multiperiod survey articles and account for almost half of the articles in the medieval and Renaissance periods. After 1600, however, this international referent declines dramatically and is replaced by specifically national references. Britain and France, already well represented in the medieval period, become the dominant referent in the two centuries following 1600, the era of the scientific and industrial revolutions. By the nineteenth century, however, all European references taken together comprise only a small percentage of articles. This is due to the overwhelming popularity of the United States as a subject after 1800. The handful of references to Russia are, in the main, references to the Soviet Union after the revolution of 1917.

Table 5 adds a nuance to TC's bias toward Western and nineteenth- or twentieth-century technology. We find a geographical-temporal pattern moving in a single direction from ancient technologies in the Middle East, through the classical, medieval, and Renaissance tech-

Table 4
Technologies referred to in *Technology and Culture* articles

Metals	20
Weapons (conventional)	20
Chemicals	19
Land transport	17
Production systems	16
Agricultural innovations	15
Civil engineering	15
Water transport	15
Tools and instruments	12
Rockets	11
Steam engine	11
Internal combustion engine	11
Hydropower	10
Agricultural technology	9
Architecture	9
Air transport	9
Machine tools	9
Automata, clocks	8
Urban engineering	7
Textiles	7
Electrical technology	7
Weapons (nuclear)	6
Electric generation and transmission	4
Fuel	5
Mechanical power transmission	4
Computer	3
Telegraph	3
Telephone	3
Environmental technology	3
Mechanical feedback mechanisms	2
Musical instruments	2
Paper	2
Ceramics	1
Telemetry	1
Photography	1
Radio	1
Medical technology	1
Miscellaneous	65

italicized subcategories refer to one specific technology

Table 5
Correlation of time periods and core place references

	Ancient	Classical	Medieval-Renaissance	Scientific and Industrial Revolutions	Nineteenth Century	Twentieth Century	Several periods	Totals
United States				3	52	53	6	114
Europe	1		14	5	2	1	10	33
Britain and France			8	12	15	2	2	39
Germany			1			5	2	8
Russia					2	5		7
Other European			5	1	1	4	1	12
Greece and Rome		9						9
Middle East	5	1						6
Asia	1		2		5			8
Latin America					1	2	2	5
Africa								0
Australia								0
Totals	7	10	30	21	78	72	23	241

nology of the Mediterranean basin and Europe, to the "scientific" and "industrial revolutions" situated particularly in Britain and France, and finally into the contemporary technology of the United States.[71] As a community the authors tend to perceive the United States as the cutting edge of technology in the nineteenth and twentieth centuries and to ignore recent European technology almost as completely as they do all non-Western technology in any era. TC is an American journal, and the bias is to some extent natural. Nevertheless, the complete commitment to "the West" as *the* place of technology and the one-directional pattern just mentioned are evidence of a substantial cultural bias in the body of articles.

Apart from this bias, the absence of significant patterns is the most revealing characteristic of the map of the historical terrain found in TC. We find a map of an uncharted wilderness or, at best, a frontier. While it is true that most contributing scholars selected their sites within the Western unilinear bias, it is also clear that the historical terrain within that broad set of boundaries is so unmapped by prior scholarship that each new research site could be selected with relatively little consideration for preexisting canons of temporal, geographical, or topical orthodoxy. More than anything else, the three-dimensional map of site boundaries reveals the youth of the history of technology as a distinct discipline.

When the profile of article references across the time periods of Western civilization is correlated with the five methodological styles discussed above, we find several patterns that raise intriguing questions about the temporal and geographical predilections of contextual, internalist, and externalist historians of technology. Before looking at the data of table 6, let us recall the major conclusions about the three historical styles in TC. We noted above that the internalist and contextual styles are based on strikingly different methodological foci—the internalist's use of technological design as the organizing principle for research and the contextualist's use of the tension between design and ambience for the same purpose. The externalist style was seen to be dependent on a tradition of contextual history. Not only did it develop in the later years of TC's publication, but its primary organizational principle is the technological ambience itself. We noted that the concept of a technological ambience results from the cumulative effect of contextual research. It could be argued, therefore, that time periods dominated by contextual articles should also show a healthy percentage of externalist articles and that periods dominated by internalist history would be those in which the tension between technical design and the specifics of its cultural ambience had not generated

Table 6
Correlation of time periods and methodological styles

	Ancient	Classical	Medieval-Renaissance	Scientific and Industrial Revolutions	Nineteenth century	Twentieth century	Several periods	No time references	Totals
Contextual		4 (40%)	7 (27%)	14 (67%)	52 (74%)	43 (57%)	15 (39%)	1 (4%)	136 (50%)
Internalist	5 (83%)	6 (60%)	16 (62%)	3 (14%)	3 (4%)	6 (8%)	8 (21%)		47 (17%)
Externalist			2 (8%)	4 (19%)	13 (19%)	13 (17%)	5 (13%)		37 (14%)
Nonhistorical	1 (17%)					6 (8%)	5 (13%)	20 (77%)	32 (12%)
Historiographical			1 (4%)		2 (3%)	7 (9%)	5 (13%)	5 (19%)	20 (7%)
Totals	6	10	26	21	70	75	38	26	272

great interest among contextual historians. With these hypotheses in mind, it is enlightening to look at table 6's correlation of the three styles with the various time periods.

The internalist style clearly dominates the ancient and medieval-Renaissance periods. The classical period is marked by a more even division between contextual and internalist research. The four contextual studies indicate that their authors considered the Greek and Jewish cultural ambiences to be influential for the design of the technologies in question.[72] On the other hand, the contextual dominance of the periods from 1600 to 1900 is extraordinary and contrasts sharply with the minuscule proportion of internalist articles in the same two periods. It is clear that the tension between technical design and its ambience in these three centuries is of great interest to TC scholars. Not surprisingly, we find that externalist studies are almost nonexistent in the pre-1600 periods. On the other hand, they comprise a healthy 20 percent of the articles covering the periods from 1600 through 1900. The twentieth century reveals a unique pattern. Although contextual and externalist articles are still the most numerous, their dominance is muted by the presence of three equal-sized groups of internalist, nonhistorical, and historiographical articles.[73] The small size of our sample of articles does not permit us to draw firm conclusions from these patterns, but they do enhance our understanding of the characteristics of the three historical styles as they appear in TC. These patterns confirm our previous finding that the primary contrast, in terms of methodological style, is between internalist and contextual history, with externalist history following the pattern set by the contextualist writings.

Significant convergences in the geographical, temporal, and topical research sites of the articles could be a source of shared discourse for historians of technology. The evidence reveals that this has not been the case in SHOT's first two decades. It might be suggested that it is simply too early in the life of the discipline to find any significant shared discourse, but this is not so. As we shall see in the next three chapters, the themes that TC authors have adopted to interpret the history of technology provide evidence of some well-developed shared discourse. This is all the more remarkable in the light of the diffuse pattern of site selection found here.

2

Emerging Technology and the Mystery of Creativity

There is, indeed, no reason to believe that technological creativity is unitary. The unknown Syrian who, in the first century B.C., first blew glass was doing something vastly different from his contemporary who was building the first water-powered mill. For all we now know, the kinds of ability required for these two great innovations are as different as those of Picasso and Einstein would seem to be.

The new school of physical anthropologists who maintain that Homo *is* sapiens *because he is* faber, *that his biological differentiation from the other primates is best understood in relation to tool making, are doubtless exaggerating a provocative thesis.* Homo *is also* ludens, orans, *and much else. But if technology is defined as the systematic modification of the physical environment for human ends, it follows that a more exact understanding of technological innovation is essential for our self-knowledge.*

—Lynn White, Jr.[1]

With these words—the conclusion of his paper at the 1962 Encyclopaedia Britannica Conference on "The Technological Order"—Lynn White, Jr., captures much of the perplexity and wonder that continues to surround the mystery of technological creativity. How does it happen that human beings create structures previously unimagined? How does human imagination interact with the complexities of the technological status quo? And finally, how do these technical interactions—new designs with old—relate to the tangled web of culture and society? White spoke for many members of SHOT with his observation that "a more exact understanding of technological innovation is essential for our self-knowledge." This question, more than any other, absorbed the attention of TC's authors during the journal's first two decades, and it is no surprise that they form the foundation for the thematic analysis of this and the following two chapters.

Historical research is communal. It is a creative endeavor, but the creativity is a shared venture and it is rooted in the cultural biases of the community of historians. From this central premise, articulated in chapter 1, flows the structure of this book. It can be argued, from the steady, if unspectacular, growth of SHOT and TC and from the perceptions of TC authors, that the society and its journal have begun to create a forum in which new historical language about technology is developing. As noted in chapter 1, three areas of decision making—site selection, priorities among types of evidence, and thematic interpretations—give us frames of reference for reading TC's articles in search of shared discourse. TC's evidentiary priorities reveal that contextual history has become the dominant methodological style in the SHOT community, with internalist and externalist styles as contributing bodies of research. A survey of trends in the temporal, geographical, and topical boundaries of research sites indicates a strong Western bias but little else.

These findings have proven helpful for understanding the cognitive structure of TC's historical styles, but they do not tell us much about the contents of the articles. For chapters 2 through 4, the question of content is pivotal. Two hundred seventy-two articles contain many assertions of fact. Were we to list them, our only achievement would be to demonstrate the bewildering variety of subjects that can be discussed under the rubric of "technology and culture." Historical research is, of course, more than a fact list. Its central characteristic is the creation of themes that interpret evidence. The historian must ask: "Why is this data interesting?" "What does it mean?" By creating hypotheses responding to these questions, the historian transforms raw data into a coherent story. A *community* of historians who have begun to participate in a single forum should begin, over time, to generate a number of communal themes. If none were found in TC's first twenty-one years, it would be difficult to argue that the authors were engaged in meaningful discourse. In this and the next two chapters we will present a variety of themes that have been discussed by significant numbers of TC authors. The analysis of the themes serves two purposes: they provide the best available evidence of the present level of shared discourse in TC, and they enable us to reflect, in some detail, on the language with which the historians here interpret technology.

The precise definition of the term "theme" as used here is important. By "theme" I mean a composite, hierarchical, and asymmetrical concept, created by author usage, which functions as an organizing principle for a number of more specific interpretative hypotheses related to it.

A theme is composite because no one author's hypotheses articulate all its aspects. It is hierarchically structured because the theme is so broad that it includes distinct subcategories. Any subcategory can itself be further subdivided, if usage demands it, at greater levels of specification.[2] Because the articulation of the theme into subcategories is determined by usage rather than according to an abstract definition, the hierarchical structure tends to be asymmetrical. Thus, for example, if only three articles adopt the potentially rich thematic concept of Marxist technological dynamism—this is, in fact, the case—it would distort the image of TC's shared discourse to locate "Marxism" as a broad cover term with many subcategories under the general theme of "emerging technology." Even though such subcategories could be identified in the abstract, they are not operative in TC to any significant degree.[3]

The three major themes—emerging technology, characteristics of technological knowledge, and technology and culture—the subjects of chapters 2 through 4, are complex and interconnected. Although it is impossible to separate them completely, it is possible to distinguish among them and the division provides a helpful matrix for presenting the many individual hypotheses advanced in the articles.

The format for presenting emerging technology in this chapter will be followed in chapters 3 and 4 as well. After an introductory overview each subcategory found in TC usage will be discussed in detail. The fact that a number of articles discuss the same subcategory does not mean that all adopt the same point of view. When helpful, therefore, I will use quotations from specific articles to highlight contrasting approaches.

An Overview of the Theme

In the single most explicit article about emerging technology, Lynn White, Jr., reflects on the complexity of the question. Note that his expression "the act of innovation" is an inclusive term parallel to our cover term emerging technology.

> In terms of eleven specific technological acts, or sequences of acts, we have been pondering an abstraction, the act of innovation. It is quite possible that there is no such thing to ponder. *The analysis of the nature of creativity is one of the chief intellectual commitments of our age.* Just as the old unitary concept of "intelligence" is giving way to the notion that the individual's mental capacity consists of a large cluster of various and varying factors mutually affecting each other, so *"creativity" may well be a lot of things and not one thing.*;[4] (my italics)

Table 7
Emerging technology: subcategories and frequency of reference

	All references	Impact on new technology	Impact of new technology
Process			
Invention	65		
Development	54		
Innovation	57		
Ambience			
Technological support network		63	24
Technical tradition		60	11
System		69	4

White's point is well taken; as an interpretative theme in the articles, emerging technology is both pervasive and elusive (appendix 3.1–6). Only fifty articles (18 percent) exclude it from consideration, but the status of the theme in the remainder is complicated by the variety of approaches taken to it.

Emerging technology is the central focus of some articles, but more commonly it operates either as one of several elements in a composite thesis or as a nearly unconscious presupposition influencing the author's general approach. White's "The Act of Invention" stands alone as an attempt at comprehensive analysis of the phenomenon, and we have seen that he concludes that a single definition is impossible. Over TC's first twenty-one years, however, the processes of shared discourse have begun to generate a recognizable language for interpreting emerging technology. Recently discussion of the theme has begun to take the form of explicit interaction and mutual critique. This discourse is based on the accumulation of many previous articles, which tend to discuss the topic without mutual interaction. When we take all of TC's approaches to emerging technology, it is possible to identify six points of view as the structure for our analysis. Since the comparative popularity of these six subcategories provides a profile of TC's treatment of the theme as a whole, the following overview is prefaced by table 7 indicating the number of articles interpreting emerging technology in each subcategory.[5]

In the language of the articles we find frequent reference to three dimensions of the process characteristic of emerging technology. Following common usage, they have been named "invention," "development," and "innovation." The process is almost always seen to be occurring within some ambience. In TC usage the ambience of emerging technology includes technological, scientific, and cultural factors. The technological ambience is approached in three ways, which can be identified as "technological support network," "technical tradition," and "system." As noted, the scientific and cultural factors are the subjects of later chapters. This chapter will discuss these six distinct approaches to emerging technology.

"Invention," which is often seen as the first stage of emerging technology, is an essentially personal act, shrouded in the mystery of the individual inventor. By contrast, "development" is almost always seen as a group endeavor. The term has come to mean a series of goal-directed experiments by which an inventive idea is transformed from an abstract possibility into a physical reality capable of functioning in the world of existing technologies. "Innovation" refers to the entrepreneurial activities by which the new technology is brought into the functioning technological world.

The "technological support network" of a given nation or region refers to some combination of existing processes and tools, skilled persons, or bodies of technical knowledge directly related to the new technology in question. It can be distinguished from the concept of a single "technical tradition" in the following way. A single tradition is understood to extend through a number of time periods during which the technology is refined by an increasingly focused body of expertise. By contrast, technological support network refers to an array of technical traditions but only as they constitute the technological ambience of a single time and place.

A third approach to the ambience of emerging technology is systemic. Whether the system is as simple as a single machine or as complex as an entire production process, the "system" concept calls attention to the universal technological necessity for functional integration. Put simply, the component parts of any system must fit one another if the whole system is to function successfully. As we shall see, this approach differs considerably from both the technological support network and technical tradition.

TC's usage suggests another distinction pertaining to these three descriptions of ambience. In some articles the ambience is understood as existing *before* and exerting influence upon the new technology in question. Other articles refer to the ambience as existing *after* and being influenced by the new technology. The two approaches are not equally balanced. It is much less common to discuss the impact of a new technology on its ambience than the other way round.

The Process of Technological Change

The three terms most commonly identified with the process of technological change in TC are invention, development, and innovation. Several articles have consciously integrated them within a tripartite model defining them as three intellectually discrete dimensions of the process. Other articles, while equally explicit in their definition of the terms, restrict their attention to one or two of the three. Finally, many articles describe the same phenomena without using these terms. Before considering them as a tripartite model, we will discuss the intellectual profile of each dimension individually.

The act of invention

In his *Trattato di architectura*, the fifteenth-century Sienese inventor and painter Francesco de Giorgio Martini articulated the inventor's perennial complaint.

> I am reluctant to show them [my inventions] forth to all, for once an invention is made known not much of a secret is left. But even this would be a lesser evil if a greater did not follow. The worst is that ignoramuses adorn themselves with the labours of others and usurp the glory of an invention that is not theirs. . . . If in all epochs this vice hath abounded, in our own it is more widespread than in any other.[6]

Given the lamentations of inventors over the centuries, Martini's claim to an epoch preeminent for rudeness to inventors might be called into question. What cannot be disputed, however, is the passion surrounding the delicate task of awarding recognition and financial remuneration to the person whose insight has created a new technological design. Accurate verification of the true inventor has been a preoccupation in the West for centuries and has given rise not only to the tangled complexities of patent law, but also to a large body of literature in the history of technology assessing the validity of claims to originality. It is not surprising that a large group of TC articles, thirty-five in all, adopt this approach to invention (appendix 3.1.1). Most limit themselves to an attempt to verify or discredit a given claim. Thus in the article about Martini, Ladislao Reti traces a number of instances of plagiarism that substantiate Martini's complaint.

Identifying the true inventor is not always the simple matter of establishing who first conceived of a given design concept. It can be argued that an individual "invents" a new concept only when s/he recognizes its usefulness and succeeds in communicating its importance to an appropriate audience. Thus Lynwood Bryant rejects the argument that Beau de Rochas's prior articulation of the four-stroke cycle constitutes a valid refutation of Nicolaus Otto's claim as the true inventor.

> Beau de Rochas undoubtedly conceived the idea of the four-stroke cycle, but he did not recognize its significance, and neither did anybody else at the time. If one is concerned about credit for the invention, then the appraisal of Beau de Rochas' claim is a matter of determining *what is meant by invention* and what sort of credit should be awarded for a *pure conception without realization or influence*. To a historian tracing the evolution of the internal-combustion engine, this early expression of a key idea is very interesting; *but it is a dead-end idea* which had no influence on the development of engines. The idea had to be reconceived by Otto.[7] (my italics)

For Bryant, the historically significant invention is more than a novel concept; it is also an act of communication, an articulation of the concept which has historical consequences.

Other historians show an awareness of the communicative dimension of invention when they focus attention on the importance of specific language of patent claims (appendix 3.1.2). James E. Brittain and Eric Robinson both argue that the ability of the inventor to satisfy patent offices, that is, to communicate effectively with them, is a critical factor in achieving credit as the true inventor. Brittain argues that the insistence of W. W. Swan, patent lawyer for American Bell Telephone, that Cambell's patent "not contain any mathematical formulas" weighed heavily against his claim and in favor of Pupin's. In fact, Swan's inability to understand Campbell's loading graphs delayed the entire filing process. Robinson's article takes as its central focus the linguistic complexities of James Watt's patents and subsequent litigation concerning them.[8] Thus the evaluation of claims to originality, which at first glance appears to be a straightforward historical task, raises the question of the nature of the act of invention itself.

Invention is always the act of an individual person who responds to the existing state of the art by creating a new insight. TC's historians rarely approach the topic of the individual creative process, however, and those who do so only point out its mysterious nature (appendix 3.1.3). Lynn White, Jr., after arguing the probability that the plank boat design originated with the Chumash Indians of the Pacific Coast, raises the question of the original inventor.

> Since a group can conceive of nothing which is not first conceived by a person, we are left with the hypothesis of a genius: a Chumash Indian who at some unknown date achieved a break-away from log

> dugout and reed balsa to the plank boat. But the idea of "genius" is itself an ideological artifact of the age of the Renaissance when painters, sculptors, and architects were trying to raise their social status above that of craftsmen. Does the notion of genius "explain" Chumas plank boats? On the contrary, *it would seem to be no more than a traditionally acceptable way of labeling the great Chumash innovation as unintelligible.*[9] (my italics)

There is no historical interpretation of the precise nature of invention in TC. Joseph Agassi's "The Confusion between Science and Technology in the Standard Philosophies of Science" is the only explicit attempt to analyze the psychological process of inventive thinking. He is, of course, a philosopher rather than a historian.[10] Articles exploring the nature of invention in a thematic fashion tend toward one of two general approaches. Some study individual inventors in terms of personality or intellectual background. Others reflect on the function of the creative insight within the process of emerging technology. We will discuss the two approaches in that order.

Many articles treating the personality of inventors focus attention on the motives engendering inventive behavior (appendix 3.1.4). Otto Mayr describes the personality of Charles Porter, inventor of the high-speed reciprocal steam engine, in terms of motivation.

> Charles T. Porter was an American engineer of the heroic mold, self-taught, unorthodox, and independent, who delighted in boasting about his innocence in matters of theory. . . . His ventures into theory therefore present some intriguing ironies. It turns out that, in spite of his pose as an "eminently practical" man, he undertook those ventures mostly for *an ancient but non-utilitarian motive*, namely, *intellectual curiosity, the innate urge to solve riddles.*[11] (my italics)

Throughout the article Mayr argues that Porter's curiosity was the source of a series of insights leading to his theoretical analysis of high-speed engines. In similar fashion, David Hounshell suggests that the desire to reap financial reward was one of several important dimensions in the inventive energy and focus of Alexander Graham Bell and his competitor, Elisha Gray. Cyril Stanley Smith adopts a different approach when he argues that invention often emerges from an aesthetic context. Sir Robert Watson-Watt asserts that the primary motive for technological creativity is the humanitarian desire to ensure "the fuller enjoyment of life . . . living more abundantly." Finally, in his reflection on possible stimuli for Eilmer of Malmesbury's attempt at heavier-than-air flight in the eleventh century, Lynn White, Jr., argues that the monk's religious faith may well have been the source of his remarkable attempt. He bases the argument on evidence suggesting that "Anglo-Saxon England in his time provided an atmosphere conducive to originality, perhaps particularly in technology." The motivational factor at the heart of this cultural atmosphere was religious, a new understanding of the nature of God who, in the eleventh-century Winchester Gospel Book is portrayed iconographically as "the master craftsman . . . holding both scales and a pair of compasses."[12]

In these examples five motives are suggested as sources of inventive behavior: curiosity, profit, love of beauty, benevolence, and belief in God. Such personal motives are never presented as a complete explanation of the relationship between the personality of the inventor and the act of inventive insight itself. The lack of any thematic interpretation of this relationship and, indeed, the lack of consensus about which motives are most pertinent to invention, are further evidence that TC's authors have not developed any interpretative model linking personality and inventing.

Another group of articles focuses on the intellectual background of the inventor (appendix 3.1.5). They see the inventor's intellectual perspective as a necessary condition for insight rather than a direct cause. Robert Fox's interpretation of the "chance" observation leading to the invention of the fire piston exemplifies the approach.

> Clearly the fact that [the fire piston's] invention was . . . the direct consequence of a *chance observation* rather than a deliberate attempt to apply newly acquired scientific knowledge to the long-established problem of fire-making, must diminish to a certain extent the importance of this scientific background. *But the background cannot be ignored completely*. Without it the ignition (by the rapid compression of air) of a small piece of linen in the exit tube of the condensing pump . . . might well have been *overlooked or at least misinterpreted*.[13] (my italics)

According to Fox, the function of the scientific background is indicated by the word "background" itself. Knowledge, whether scientific as here or an understanding of the functioning of the business world as in Reese V. Jenkins's study of George Eastman, is not the origin of an inventive concept but the personal and intellectual frame of reference that is a necessary prerequisite for every inventor if s/he is to interact with some aspect of the state of the art.[14]

The second general approach to invention is focused not on the inventor, nor on the act of insight, but on the resulting concept itself. This approach can, in turn, be divided into two types. The two differ on only one point and even here the distinction is often blurred. Is the inventive concept best understood as the first stage in the process of emerging technology or as a distinctive type of intellectual activity pervading the process? One insight is common to both approaches. The inventive concept is understood to be an abstraction that must be embodied through processes often called "development" or "innovation" before it can take its place in the existing technical ambience.

The suggestion that "invention" refers to the first stage in the process of emerging technology (appendix 3.1.6) is more a linguistic implication than an explicit position. Thus for example:

A published patent reveals the date of application. It does not, however, reveal either that of conception or, more important, that of a *beginning* of reduction to actual practice.[15] (my italics)

Again:

Earlier inventions had come about, it has been said, through the marriage of an inventive idea with the means for its realization; *the idea came first, a search for means followed.*[16] (my italics)

Neither text necessarily argues that inventive insight functions *only* as the origin of a process, but the language implies that a criterion for calling an idea "invention" is its role in starting a process of emerging technology.

The second approach stresses a different aspect of inventive insights (appendix 3.1.7). Not all such insights function as an originating concept governing an entire process. If the term "invention" is to refer to any act of insight resulting in a new design concept, then it is easy to see that the entire process of emerging technology is studded with such inventions. The principle is articulated in some detail by Lynwood Bryant in his study of Rudolph Diesel.

Diesel's activity during this period could be classified as either invention or development. Looking back later, Diesel regarded it as invention. An invention, he said, is not a pure idea, but rather the product of a struggle between Idea and Nature. The product is always a composite between the Ideal and the Attainable, and working out the details of this compromise is a part of the process of invention. To me this work looks more like what I would call development: the messy job of constructing something new and trying to make it work . . . coping with unexpected difficulties, retreating from one approach and trying a *new* one . . . that sort of thing.

At the same time, every day, Diesel was deeply involved in *theoretical* work. . . . The real engine had to be rationalized. An invention as described in a patent may be the embodiment of a *permanent, fixed idea*, but in real life *ideas* may not stay fixed.[17] (my italics)

Bryant's point is clear. If "invention" must mean the fixed and permanent idea described in the patent, then some other term must be found for the other creative insights occurring throughout the process.

The tension between "invention" as originating concept and as the conceptualizing activity pervading the process of emerging technology may well be resolved by distinguishing between marginal and breakthrough inventions. A breakthrough invention is a key idea that originates and governs the process of emerging technology. A marginal invention is an insight providing some incremental change within the process itself. This distinction pervades much of TC's discussion of emerging technology. As we shall see below, it is not limited to the treatment of inventions. In fact, the distinction is most often discussed in relation to innovation, technical tradition, and systems.

The process of development

"Development," the second dimension of TC's interpretation of emerging technology, is normally understood as an intermediate stage between the act of invention and the introduction of an innovation into normal use (appendix 3.2). In the words of Walter Dornberger, development is "a matter of progressing step by step . . . weighing the feasible against the hoped-for."[18] Many TC historians have chosen to study situations where people have wrestled with the tension between what is "feasible" and what had been "hoped-for" in the original inventive idea. They show remarkable consistency in their understanding of what is involved in the process. "Development" seems to have entered common usage toward the end of the first decade of SHOT's existence. Even when the word is not used, however, it is clear that, with few exceptions, authors define the process by the same characteristics.

Development references are nearly unanimous in assuming that the process includes two interrelated activities, the construction and testing of models (appendix 3.2.1). While many articles describe these activities without interpretation, a few are more explicit. Thus, F. M. Scherer summarizes development as a process of "building, testing, modifying, and retesting models of increasing scale and sophistication."[19] In his article about the diesel engine, Lynwood Bryant paraphrases machine designer Alois Riedler's assessment of the developmental work at Maschinenfabrik Augsburg from 1897 through 1902 in the following manner.

> [Riedler] divided the process of developing the diesel (or anything else) into three stages. First, he said, is the development of a machine that works, an engine that runs (that is *gangbar*). The second stage is the development of an engine that is useful (*brauchbar*), that can carry a load, reliably, for a reasonable time. That is quite a different thing from the *gangbar* engine, he said, and it may be a greater and more costly achievement. But this is not the end of development; a third stage, the most important of all, is necessary to make the engine *markfahig*, that is, ready to be sold, able to take its place and hold its own in the existing economic system.[20]

The difficulties inherent in the process described by Riedler stem from the tension between the abstract ideal of the inventive design and the concrete limitations encountered in the development process. At every stage of Riedler's process a functioning model of the diesel engine concept had to be constructed with the machines, materials, and skilled personnel available to Maschinenfabrik Augsburg.

An equally important constraint, in this as in every developmental project, is the cost of the lengthy process. It is not surprising that nine articles about development stress the importance of securing funds to support the work (appendix 3.2.2).[21] The importance of adequate funding also helps explain why most projects treated in TC are situated in a military or a business context, either in passing or as a central focus (appendixes 3.2.3–4). Thus Stewart Leslie's study of Boss Kettering's development of the copper-cooled engine at General Motors in the early twenties explains the project's failure by a variety of interdivision tensions and conflicting market perceptions within the firm. John H. Perkins treats the influence of the military establishment on shifting priorities in entomological research and insect-control practices during World War II. In their detailed attention to the bureaucratic, political, and institutional dimensions of development, both are typical of many journal articles.[22]

Inherent in the dual activities of model building and testing is a second major characteristic, goal directedness (appendix 3.2.5). Goal directedness is a subtle business, involving as it does the peculiar character of information flowlines in developmental dynamics. The implications of development's goal directedness become clearer with the help of an analogy from engineering, the concept of negative feedback.

Negative feedback can be defined as a functioning system that includes an error-reading device. When the system reaches a preset level of operation, an error signal is triggered and is fed back through an information loop to correct the system. A simple example is the heating system of a building. When the building is being heated it must, in principle, overheat past the preset temperature of the thermostat. When the temperature reaches the thermostat setting, the device "reads an error" back to the furnace and shuts it down. Although the signal is triggered when the temperature reaches the preset level, it is correctly called an error signal because the momentum of the furnace will necessarily carry the temperature past that level before the corrective action takes hold.[23] In similar fashion, it can be said that the physical construction of each test model creates a "momentum" of its own. Momentum is a physical reality and here the term refers to the fixed, physical characteristics of the model as built. Some of those characteristics will prove to be "errors." Testing the model is, of course, intended to identify these errors. The shortest distance between the two points of an invention and its final embodiment in a viable innovation is not a straight line but rather the back-and-forth dynamic of trial, error, correction, and retrial.[24]

The negative feedback metaphor focuses attention on a critical dimension of the developmental process which has not escaped the attention of TC's authors. If the process of testing and retesting models depends on the ability to recognize error and feed that information back into the design of subsequent models, then it follows that the entire process must be what is called a "closed" rather than an "open" system. A closed system is characterized by a circular information loop from feedback device to power source. But development would appear to be an open-ended process. It is axiomatic that the final embodiment of such a process is unknown at the beginning. What then allows development to function with the negative feedback characteristic of a closed system?

It is here that the goal-directed quality of development comes into play. Goals create a coherent cognitive system, which, by defining the goal in specific terms, makes it possible to determine what "success" and "failure" mean for the project. It would appear that this goal directedness is of greater thematic interest to TC's authors than are model building and testing. Many references to modeling and testing are simply descriptive, and indeed fifteen articles treat development by listing "solved" problems with no further comment (appendix 3.2.6). By contrast, twenty-five articles contain explicit thematic analysis of the process of defining a project's goal.

Robert L. Perry is typical of a handful of historians who treat recent "high technology" development projects and who stress the necessity of an authoritative team design. Leadership is especially important for redefining the goal direction in light of ongoing test results. Writing of the ICBM team that developed the Atlas missile, he notes that "authority and responsibility had to be consolidated in one agency, else management control would be neither quick reacting nor responsible." Perhaps overlooking Leslie Groves and the Manhattan Project, he describes the "unprecedented authority for program management" as an important dimension of the Atlas Project.[25]

The impact of authority on project goal definition is refined by R. Cargill Hall's contrast between the satellite project design prepared by North American Aviation for the navy and the distinctly different design created by Project RAND for the air force. The governing factor leading to the differences is seen to be the *stated goal* as articulated by the navy and the army air force.

The engineers of North American Aviation were aware at that early date that a satellite could be made by multi-staging conventional [V-2] rockets, but *they were responding to the Navy's requirements to see if a single-stage satellite could be made*. . . . In the Army Air Force satellite studies . . . *no predetermined specifications were imposed* on Project RAND.[26] (my italics)

Due to the navy's requirements, the development proposal submitted by North American Aviation differed radically from Project RAND's, although both teams of designers were familiar with the constraints of existing technology.

Thomas M. Smith's study of the development of a computer for Project Whirlwind is another striking example of the importance of the goal-defining role of military authorities. The shift from the "engineering-development-oriented staff" of the Naval Special Devices Division to the "research-science-oriented" staff of the Office of Naval Research—which supplanted the NSD in 1946—was a shift of "fundamentally differing design philosophies" that nearly destroyed the project. The ONR staff had little sympathy for the funding requirements of Project Whirlwind, which was originally intended to provide the control system for a cockpit without wings, a flight simulator for pilot training. Only the adoption of the project by the air force for an entirely different purpose (a computer control system for the Distant Early Warning radar system) saved the project from foundering on the shoals of shifting military goal definitions.[27]

Authoritative decisions are not the only style of goal definition identified by TC historians. Thomas P. Hughes's study of developmental testing related to the corona problem in high-voltage transmission of electrical power indicates that the identification of a critical problem by Charles F. Scott during research at Westinghouse Electric and Manufacturing Company functioned as a goal definition for a substantial body of later research. It is noteworthy that in Hughes's case study the development work took place not only at Westinghouse but at other research sites. Thus it is not necessary that the closed system of goal-directed developmental activity be created by the authority structure of a single team project. In Hughes's study it is the existence of communication lines within the electrical engineering community, particularly through technical journals, which appears to be the structural mechanism for feeding back the test results into the ongoing developmental process.[28]

An added dimension of goal definition can be seen in Lynwood Bryant's intriguing observation that the entire development project leading to the four-stroke Otto engine at Gasmotorenfabrik Deutz was governed by a design concept that may have been inaccurate.

> This *idea* [stratified charge] may have been false . . . but *it led to a successful engine* because it encouraged Otto to try the engine without the free piston and to carry through the idea of the four-stroke cycle as a method for achieving the compression of a fuel within the cylinder.[29]

The accuracy of the stratified charge concept that defined the goal of Otto's developmental work was less important than the fact that it served to define the goal. Thus Bryant underscores the critical point lying at the heart of the interpretations given by Perry, Hall, Smith, and Hughes. The process of development is inconceivable outside of the intellectual parameters established by a specific goal, parameters that create the closed system necessary for the feedback involved in a trial-and-error process of model testing.

One further important distinction in discussions of development is found in the common expression "research and development." It seems clear that historians of technology have borrowed the term "development" and given it the specifically historical definition just outlined. The term "research" has not been elaborated with the same thematic precision.[30] When the distinction has been treated by TC authors, it is in the context of the remarkably confused debate over the contrasts and interactions between science and technology, which will be discussed in chapter 3.

In conclusion let us note several contrasts between invention and development. Where invention is the act of an individual, development is normally the work of a group. Inventive concepts are abstractions, and developmental thinking is rooted in concrete limitations. It is perhaps not surprising that in commonly accepted history inventors are seen as heroic figures because the creation of a radically new insight remains one of the most fascinating of all human endeavors. It may be argued, however, that the development process, immersed as it is in the primordial encounter between human insight and the truculence of the specific situation, reveals more of the unique character of technology. It is perhaps for this reason that the historians of technology in TC have developed a much more precise consensus about development than they have about invention.

Innovation

To place our discussion of innovation within the context of invention and development, it helps to contrast these three very different dimensions with the help of the design-ambient model elaborated in chapter 1. We noted there that the heart of the technological problem, for engineer and historian alike, is the tension between the design symmetry required for the internal functioning of any technological artifact and the decidedly asymmetrical complexities of the ambience in which it operates. If we assume that the two poles of this tension define technological activity, we can then identify the specific meaning of "design" and "ambience" in the three stages of emerging technology as follows. In the act of invention the design-ambient tension refers to the tension between the inventor's perception of the present state of things—of the technological and societal ambience—and his or her new design insight, which creatively changes that perceived state. Both poles of the tension exist completely within the mental universe of the inventor. In development the design pole of the tension is still the original inventive idea—modified of course, by its continual redefinition during the development process. The ambience is the development project itself complete with its specific leadership, personnel, costs, time constraints, and so forth. Finally, in the innovative process the design is the inventive concept as it has become embodied during development. It stands in tension with the ambience which is often called "the real world," the existing economic and cultural reality. Thus the ambience for innovation is no longer controlled either by the mental universe of the inventor or by the goal definition of the development project.

The three stages of emerging technology are differentiated by the increasingly complex and uncontrollable character of the ambience within which the new technical design must operate. At the level of invention one asks, "Is it a good idea?" At the level of development one asks, "Can we get this design to function within the constraints of our situation?" At the level of innovation one asks, "Will this new artifact survive in the wide world?"

The phrase "three stages" is somewhat deceptive in suggesting that invention, development, and innovation follow one another in sequential order. In fact, they occur more or less simultaneously, as Lynwood Bryant makes clear.

> The three kinds of human behaviour that we label invention, development, and innovation are going on more or less all the time in the process of technological evolution. The emergence of new ideas, which we label "invention," clearly takes place throughout the process. The refinement of design that comes with experience in the real world, which we label "development," is endless. The effort to fit a new technique into the existing economic and social structure, which we label "innovation," is a guiding force from the beginning.[31]

Perhaps the confusion of language in TC's articles on this point—we find frequent use of terms suggesting linear progression—is due to the fact that the three types of activity follow a *logical* progression toward participating in an increasingly complex and uncontrolled ambience.

Innovative activity, as treated in TC's articles, reflects the helpfulness of this threefold contrast. Those articles that reflect most deliberately on innovation define it by contrasting it with invention, and occasionally with development as well (appendix 3.3.1). Edward W. Constant II's study of the turbojet engine is a formal attempt to articulate a new hypothesis explaining what is sometimes called a "breakthrough innovation" and which he names a "major technological change." Acknowledging his debt to Thomas Kuhn's hypothesis of revolutionary scientific paradigms, he defines a "technological paradigm" as "an accepted mode of technical operation . . . as defined and accepted by a relevant community of technological practitioners." He notes that each paradigm generates a tradition of "normal technology," which can include within itself many inventions that are more accurately called incremental than breakthrough. In this context he describes a breakthrough invention.

> When a technological revolution occurs, however, the community paradigm changes. Technological revolution is defined here only in terms of a relevant community of practitioners and has no connotation of social or economic magnitude.[32]

For Constant, then, a breakthrough invention is a revolutionary paradigm because it challenges the normal understanding of how some technical activity should be done and succeeds in winning a following among relevant practitioners.

It is important to note that Constant understands the revolutionary paradigm as an invention, not an innovation. As the following text indicates, the paradigm is at the outset the creation of an individual "provocateur" and the paradigm concept is abstract.

It is men, not "forces," that cause technological revolution. It is men, often alone and not infrequently ridiculed, who perceive anomaly, formulate new paradigms, and provoke crisis. It is men, not institutions or governments, who ramrod the development of what to others may seem an unlikely candidate paradigm. . . . It is men in the tradition of an older, more heroic history who are the protagonists of that revolution.[33]

Constant goes on to critique the "widely held" theory that economic factors are "the chief dynamic force behind technological progress." He describes not only how the intellectual act of insight (the "invention") depends on its aesthetic appeal for the inventor, but also how the transformation of invention into innovation is marked by the increasing importance of economic considerations.

Economic factors may hypothetically affect paradigmatic change at four levels: the motivation of the individual provocateur, the availability of developmental funds, the adherence of the first few critical converts, and the communitywide adoption of the new paradigm. . . . While economic factors, as well as social or religious factors, may serve to direct men toward innovative endeavor, there is no evidence that *such motivation has any effect on the intellectual process* of anomaly recognition or new paradigm formation.

It is only after a candidate paradigm has taken shape in its creator's mind, it is only when that creator sets out to convert the relevant community, that *economic factors* can become directly determinant.[34] (my italics)

In Constant's analysis a revolutionary inventive concept is the act of an individual motivated by a combination of delight in the aesthetic appeal of the new idea and by the fervor of a provocateur. Innovation, on the other hand, is an activity rooted in the constraints and possibilities of economic factors. The desire for profit alone does not explain inventive creativity, but it is a powerful motive for the innovator's risks and labors in adopting the new concept and transforming it into a viable technology.

Several other articles stress the distinction between the aesthetic and personal nature of invention and the economic criteria governing innovation. Robert C. Post recounts the extraordinary efforts of Dr. Charles Page to secure governmental support for the development of his wet-cell electric locomotive at a time when the primitive state of electric technology doomed the project to failure.[35] F. M. Scherer contrasts the inventive personality of James Watt with Matthew Boulton's embodiment of the "five characteristics of the innovator" as described by economist Joseph A. Schumpeter.[36] For Lynwood Bryant, Rudolph Diesel's dream of creating "the universal rational heat engine" in 1890 differs markedly from the actual engine that began, after torturous years of developmental and marketing labors, to win itself "a modest place in the stationary power market."[37]

Other discussions of innovative activity either ignore invention or imply that economic constraints are operative throughout the process. David J. Jeremy's study of the Lowell Textile Mills adopts the hypothesis that the capital-intensive and vertically integrated structure of the mills, together with their market strategy of competing with British textiles in specific target areas, were the dominant factors leading to inventions that creatively modified the production system.[38] Jeremy's interpretation does not contradict Constant's hypothesis. Constant would point out that the inventions discussed by Jeremy are part of "normal technology" occurring within an existing technological paradigm. Jeremy is not concerned with breakthrough inventions.[39]

Whatever their differences about the economic motives of inventors, all these authors agree on the central importance of economic factors in the innovation process. In this, they reflect the overwhelming majority of TC authors who treat innovation. Almost every one of their articles speak of innovation in terms of a profit motive or a marketing strategy (appendix 3.3.2). Many stress the interrelated factors of cost analysis, quantity production, and standardization of parts as major factors in innovative activity after 1800. Reese Jenkins, for example, indicates that George Eastman's Kodak camera of 1887 was an innovation system including camera, film, and a marketing and processing system all aimed at the enormously profitable amateur market.[40] A handful of authors focuses attention on patent strategies of competing corporations as a specific form of economically motivated innovation.[41] Nearly two-thirds of all innovation references are situated within a corporate setting.

TC's discussions of development projects in a military setting are indirectly related to the question of economic constraints because the military has often been recognized as a major source of funding for new technology. Normally, however, such studies do not use the innovation concepts just discussed. "Innovation" language in TC tends to reflect its origins in economic theory where the more common categories deal with corporate and private sector market forces. An exception to this pattern appeared in the April 1978 issue. Daniel P. Jones traced the strategies of Chemical Warfare Service members after World War I as they fostered private corporations that could supply tear gas for police departments in the face of "the public's fear and revulsion for gas warfare" and the U.S. War Department's consequent prohibition of military use for the gas they had developed. Every private corporation providing tear gas to police departments in the twenties was founded by former CWS members. They were actively assisted by the service's Edgewood Arsenal with research design plans and small amounts of the gas itself.[42] Such detailed study of direct interactions between military and corporate interests is rare in TC.

While concentration on economic factors is the predominant characteristic of TC's innovation studies, a substantial number of articles adopt a second perspective. An innovation does not simply emerge into the marketplace. If it proves viable, it begins to permeate its ambience over time. Some authors adopt the term "diffusion" for the process (appendix 3.3.3).[43] Articles reflect on diffusion from several perspectives. Nathan Rosenberg notes that capital goods producers in the United States succeeded in fostering the diffusion of standardized production machinery because they developed a remarkable rapport with their customer companies.

> The [machinery producers] learned to deal with the requirements of their customers at the same time that the machinery user learned to rely heavily on the judgment and initiative of the machinery supplier. It was, in part, the relative harmony and mutual confidence of these relationships which made it possible for machinery makers to eliminate customer preferences that were technically non-essential or irrelevant and therefore to design more highly standardized machinery.[44]

This harmonious relationship resulted in the diffusion of specific, standardized machine designs throughout the infrastructure of the United States. Merritt Roe Smith's study of the milling machine in early nineteenth-century arsenals stresses the influence of governmental insistence on sharing design blueprints and the influence of skilled workers migrating from one arsenal to another as the two mechanisms fostering diffusion of milling machine innovations.[45]

Vernon Ruttan and Yujiro Hayami, in their study of the transfer of agricultural technology, argue that the worldwide diffusion of standardized seed hybrids occurs as the advantages of economies of scale overcome the advantages of flexible seed design. Other authors such as Thomas P. Hughes and Irwin Feller also discuss the complexities of the diffusion of innovative technology across national or regional boundaries. These problems are directly linked with the transfer of technology, a theme to be dicussed in detail in chapter 4.[46] We should note in conclusion that nine articles verify the diffusion of a particular innovation without thematic analysis (appendix 3.3.4). They are similar to the articles focusing on verification of the origin of a particular invention.

With this study of TC's interpretation of innovation we conclude our treatment of the process of emerging technology. Before turning to the technological ambience, however, it will be helpful to contrast invention, development, and innovation from historiographical perspective. A comparison of the linguistic origins, methodological styles, and chronological profiles of the three terms provides valuable insight into the way TC's authors have begun, by a process of shared discourse, to generate a historically rooted language about the emergence of technological novelty.

The linguistic origins of the "invention-development-innovation" model

Toward the end of the first decade of SHOT's existence a significant number of TC's historians appear to have adopted "invention," "development," and "innovation" as appropriate names for three distinct dimensions of emerging technology. Often the words are accepted without question and occasionally they are the focus of explicit discourse. Several articles study all three aspects of the process while others restrict their attention to one or two.[47] In addition, there is evidence that shared discourse among historians of technology played a critical role in sharpening the conceptual definitions of the terms and, in so doing, enhanced the thematic power of each. This contribution can best be understood if we reflect on the distinct ancestries of "invention" and "innovation" before considering the tripartite model.

"Invention" is by far the oldest of the three terms, rooted as it is in the long Western European internalist tradition of invention lists. Alex Keller argues that Giovanni Tortelli compiled the first such list in 1449. In the article titled "Horologium" in his *De orthographia* he not only uses the word "invention" but situates the inventions within a historical context. In contrast to the list of inventors found in Pliny's *Historia naturalis*, which is "devoid of any sense of history or development," Tortelli's inventions are divided into "old" and "new," an indication that he understood their history as a linear process.[48]

Lists of inventions, marked at times by plagiarism and at times by complaints from victims of the plagiarists, form a rich tradition from the Renaissance to the early nineteenth century. This body of literature, together with its parallel tradition of museums of science, inventions, and curiosities, grounded the words "inventor" and "invention" in an internalist tradition—the history of artifacts and of inventors' claims to originality. The intense development of patent law in Europe and the United States beginning around the turn of the nineteenth century refined the tradition by defining the "inventor" as the individual who was awarded legal rights to a given design concept.[49]

In the first third of the twentieth century, historian Abbott Payson Usher and sociologists S. Collum Gilfillan and William Fielding Ogburn created a new dimension for the terms by articulating elaborate models to explain the process of invention itself. As noted in chapter l, Usher's *History of Mechanical Inventions* was both a part of the internalist tradition and a creative departure from it. The influence of internalist categories also appears in the work of Gilfillan and Ogburn. It can be seen, of course, in those internalist articles in TC which take the verification of claims to originality as their central focus (appendix 3.1.1).[50]

By contrast, the term "innovation" appears to have originated in a tradition of economic analysis. Economists Joseph A. Schumpeter and Jacob Schmookler, in particular, are noteworthy for theories that analyze forces leading to innovative activity.[51] Their concern is not with the origins of inventive design concepts, nor with the historical validity of claims to originality. They reveal their professional perspective by focusing attention exclusively on the entry of new technology into the marketplace. It is clear from our analysis that innovation in TC has retained its economic perspective. The overwhelming majority of references interpret the concept in terms of market strategy, profit motive, and corporate structure (appendix 3.3.2). It is instructive to note, as well, that references to patent law in discussions of innovation are radically different from those related to invention. Innovation studies understand the patent system as a financial constraint that calls for economic patent strategies.[52]

F. M. Scherer's remarkable study of the Watt-Boulton steam engine (1965) is a provocative example of the process by which historians of technology have reshaped these two concepts. On the one hand, he examplifies the internalist and economic origins of "invention" and "innovation" by applying the Usherian model of invention and the Schumpeterian model of innovation to interpret the personal style of inventor James Watt and innovator Matthew Boulton. On the other hand, his is the first article in TC to integrate the two concepts explicitly. Even more significant, he achieves this integration by adopting the term "development" as a third dimension of a single process (the emergence of the Watt-Boulton engine). After applying Usher's model to explain Watt's inventive behavior and Schumpeter's to explain Boulton, Scherer concludes by noting that neither Usher nor Schumpeter adequately interprets a third kind of behavior, which he names "development."

> It would be presumptuous to suggest modifications in the theories of Usher and Schumpeter on the basis of a single case study. . . . Still, the evidence examined here at least suggests that Usherian invention and Schumpeterian innovation are *logically distinguishable* and that both invention and innovation . . . are necessary and complementary functions in the advance of technology.
>
> The position of the function "development," as defined in this paper, is more ambiguous. The successful accomplishment of development tasks is a concern of the *Schumpeterian innovator*. . . . Development can also be subsumed under *Usher's "critical revision."* Yet the amount of time and effort devoted to development as opposed to more insightful activities by Watt *implies the need for a less subordinate treatment*. . . . In view of this . . . one might wish to assign *development a functional status equal to that of invention* [*and*] *innovation.*[53] (my italics)

Scherer's cautious language—"it would be presumptuous," "one might wish"—exemplifies the process of thematic creativity. Taking two well-worked hypotheses as his starting point, he finds evidence in the Watt-Boulton story that calls for a new model. Neither Usher nor Schumpeter gives developmental activity the attention it deserves. Scherer appropriates and uses their models in the precise areas of their respective strengths, namely, the interpretations of invention and innovation. He creates a new model by restricting the thematic range of the two prior models and by integrating them through a new middle term, "development."

For Scherer, development achieves its full meaning only in relationship with invention and innovation. He argues that the substantial gap between the two, as they have normally been understood, is not satisfactorily interpreted merely by extending the range of either. Nathan Rosenberg makes a similar point five years later.

> As part of our Schumpeterian heritage, we will look upon the transformation of an "invention" into an "innovation" as the work of entrepreneurs. But from a technological perspective, it is much more the work of the capital goods industry. . . . In making new products and processes practicable, there is a long adjustment process during which the invention is improved, bugs ironed out, the technique modified to suit the specific needs of users, and the "tooling up" and numerous adaptations made so that the new product (process) can not only be produced but can be produced at low cost. The idea that an

invention reaches a stage of commercial profitability *first* and is then "introduced" is, as a matter of fact, simpleminded. It is during a (frequently protracted) shakedown period in early introduction that it becomes obviously worthwhile to *bother* making the improvements.[54]

Rosenberg's process "during which the invention is improved [and] bugs ironed out" is clearly developmental, although he does not use the term. He calls it "a shakedown period."

What does the historical community gain by naming this period "development"? Rosenberg's critique of a "simpleminded" description of the transition from invention to innovation hints at the value of a middle term, a term with as salient a name as the other two. If common historical parlance spoke of the "invention-development-innovation" process instead of the "invention-innovation" process, then the language itself would call attention to the critical function of modeling and testing in emerging technology. For historians of technology, even the two older expressions profit from their relationship with the new term. The very fact that their definitions become more limited—by the conceptual boundaries created by "development" as middle term—brings both inventive insight and entrepreneurial strategies into sharper focus. The adoption of development as middle term is, therefore, a substantial advance in precision for the historian's interpretation of emerging technology.

It is unlikely that Scherer is the only creator of the three-dimensional model. Evidence exists indicating that several historians of technology began to adapt the term "development" in a three-stage model at about the same time.[55] This is not surprising because "development" has long been part of engineering parlance, especially in the expression "research and development." Nevertheless, Scherer serves as an excellent example of the operation of shared discourse within the SHOT community. He explicitly acknowledges his debt to two of the scholarly traditions most influential in the early formation of SHOT. On the other hand, it is clear that the tripartite model developed by Scherer and others toward the end of SHOT's first decade exerted influence on other members of the community. Consider, for example, the contrast between Lynwood Bryant's 1966 study of Nicolaus Otto's four-stroke internal combustion engine and his 1977 study of the diesel engine. In both articles Bryant treats in considerable detail activities that we have defined as developmental. In the Otto article the term "development" is not used, nor are the distinctions between invention, development, and innovation clearly drawn. In the diesel study, however, Bryant not only adopts the tripartite model, but he makes a

Table 8
Correlation of methodological styles and references to technological process

	Invention	Development	Innovation
Contextual	44 (68%)	45 (83%)	48 (81%)
Internalist	10 (15%)	3 (5%)	4 (7%)
Externalist	3 (5%)	3 (5%)	4 (7%)
Nonhistorical	3 (5%)	2 (4%)	1 (2%)
Historiographical	5 (8%)	1 (2%)	2 (3%)

thematic contribution to it by noting that the three kinds of activity often occur simultaneously.[56]

The vitality of TC's discourse about the tripartite model is nowhere better seen than in a symposium, "The Developmental Phase of Technological Change," published in the July 1976 issue. In his introduction Thomas P. Hughes describes the evolution and current status of thematic discourse about the model.

> Because the history of technology is a recently cultivated field of scholarly activity, not many of its critical research problems have been identified. As research and reflection continue, however, *problems will emerge* and in some cases *will be identified as critical ones worthy of the attention of a number of scholars*. . . . One set of categories already widely used in discussing this process ["the nature of technological change"] includes invention, research, development, and innovation. . . . Therefore I asked historians of technology Lynwood Bryant, Richard Hewlett, Thomas Smith, and Charles Susskind, who like me had found development a central subject and conceptual problem in their studies, to participate in a program session that would raise and probe questions about the development phase of the process of technological change and perhaps *help the discipline decide if development is in fact a critical problem deserving of further study.*[57] (my italics)

Table 8 reveals further nuances in TC's use of invention, development, and innovation by correlating the article references to each with the methodological style of the same articles. All three concepts are primarily used by contextual historians. This is especially true for developmental and innovation references, 83 percent and 81 percent of which appear in contextual articles. In treatments of invention, where we would expect to find a strong influence from internalist history of technology, we see a significantly higher share of internalist articles. Nevertheless the 15 percent of invention references found in internalist articles is dwarfed by the 68 percent found in contextual articles. It appears, therefore, that all three concepts have been ap-

Table 9
Correlation of time periods and references to technological process

	Invention	Development	Innovation
Ancient	3 (5%)		
Classical			
Medieval-Renaissance	10 (15%)		3 (5%)
Scientific and Industrial Revolutions	7 (11%)	6 (11%)	5 (9%)
Nineteenth century	19 (29%)	14 (26%)	22 (39%)
Twentieth century	11 (17%)	29 (54%)	20 (35%)
Several periods	9 (14%)	5 (9%)	4 (7%)
No time reference	6 (9%)		3 (5%)

propriated by contextual historians for use in their interpretations of emerging technology. We have already noted in chapter 1 that contextual history achieved a position as the governing historical style in the journal as early as 1964. Its extraordinary domination of the usage of these three terms is further confirmation of that fact.

The distinct content and scholarly ancestry of the three concepts is underscored in table 9 by the contrasting temporal profiles of each set of references. References to development (the creation of contextual historians) and to innovation (originating in economic theory) are found almost exclusively in articles treating time periods after 1600. By contrast, the references to invention are broadly scattered across the entire chronological spectrum. Neither pattern is surprising when we recall that internalist articles comprise the great majority of TC articles before 1600 and that an even greater proportion of post-1600 articles are contextual. It is also noteworthy that a significant number of invention and innovation references occur in nonhistorical articles. Again, by contrast, there are no developmental references in such articles. This further confirms our hypothesis that development entered TC's discourse as a creation of the journal's contextual historians. Invention and innovation were well-defined concepts before SHOT's origin so that both were available to the authors of nonhistorical essays.

Invention, development, and innovation: in TC's universe these three concepts have come together with striking vitality into the most richly textured theme we shall find in this study. Still, they do not exhaust TC's interpretations of emerging technology. Thus far our attention has focused on the technological processes by which new technology comes into being. They do not, of course, occur in a vacuum. Hints about the nature of the ambience in which technologies emerge have been scattered throughout. We have seen references to sociocultural ambient factors such as military policy, inventive motivations, scientific traditions, and economic constraints. They will be considered in greater detail in subsequent chapters. Before turning to them, however, there remain a host of TC references to the specifically technological aspects of the ambience of emerging technologies.

Emerging Technology and Its Technological Ambience

In the overview at the beginning of the chapter we noted that TC's articles show a strong bias toward understanding the ambience of emerging technology as existing before rather than after the emergence of a new technology (192 and 39 references, respectively). This section will reflect TC's bias by paying more attention to the influence of the preexisting ambience than to the impact of new technology on its later context. The technological dimensions of the ambience of emerging technology can be divided into three types, which I have named "technological support network," technical tradition," and "systems."

The technological support network

The expression "technological support network" is meant to include the many technical capacities of a given society, capacities embodied in hardware or in the skills and knowledge of its people. The specific expression never occurs in TC. It has been created as a cover term for the many references to the relationship between emerging technology and the existing state of the art. Journal articles take several approaches to that relationship.

The most common reference to the technological support network is the explicit or implicit assertion that a given dimension of technical capacity is a necessary condition for the emergence of a new technology (appendix 3.4.1). Robert L. Heilbroner articulates the principle.

> To make a steam engine . . . requires not only some knowledge of the elastic properties of steam but the ability to cast iron cylinders of considerable dimensions with tolerable accuracy. It is one thing to produce a single steam-machine as an expensive toy such as the machine depicted by Hero, and another to produce a machine that will produce power economically and effectively. The difficulties experienced by Watt and Boulton in achieving a fit of piston to cylinder illustrate the problems of creating a technology, in contrast with a single machine.[58]

Heilbroner uses the steam engine's relationship with requisite technical capacity as an example of the general principle. Such general statements are not unusual in TC, but the relationship more commonly appears in an implicit form within the author's argumentation.

In the following passage Donald S. L. Cardwell implies precisely the same point as Heilbroner but without specifically calling attention to it as a general law.

At the Great Exhibition of 1851 only six "electro-magnetic engines" were displayed, and they were little more than toys, classified under "Philosophical, Horological, and Musical Instruments." This should be compared with the 114 steam engines which were anything but toys. *The revival of electric power had to wait* for such developments as the invention of the self-excited dynamo, the application of electric light to lighthouses and then to streets and squares, and finally to the masterly invention of the incandescent lamp by Edison and Swan.[59] (my italics)

Walter Dornberger's description of the V-2 project at Peenemünde stresses the critical need for skilled personnel and the difficulties created by Germany's allocation priorities.

Why were those four thousand soldiers at the Research and Development Center? The reason was lack of support of the rocket program by the Armament Ministry and by Hitler himself. During the war, both were responsible for manpower and raw material distribution. Although the experimental station in Peenemünde was fully supported by our military chiefs . . . we could get *no personnel and no material* from the high offices of the Reich.[60] (my italics)

The pattern in these three articles can be found in a multitude of others. Whether the technological capacity in question is a set of contributing technologies, as with Heilbroner and Cardwell, or the availability of materials and personnel, as with Dornberger, TC authors recognize that new technology necessarily depends on the specific conditions of the state of the art.

Very rarely, technological interdependence is discussed in language suggesting that the support network is more than a necessary condition for new technology; it is seen as a causal dynamic *generating* rather than *permitting* innovation (appendix 3.4.2). Writing of the ongoing emergence of an array of prime movers (e.g., waterwheels and windmills) Abbott Payson Usher uses language implying that the interrelated dynamic of prime-mover technology, in consort with its applications in an industrializing society, is the cause of further prime-mover innovations.

Full generalization in the use of power requires an array of prime movers capable of providing power of any desired magnitude, with any desirable characteristic in respect of compactness or rate of revolution. Whatever the importance of particular inventions, such a long *sequence of development* cannot be understood unless *the process* is seen as *a progressive and convergent synthesis*. . . . The *complete array of prime movers required* the further development of steam in the non-condensing or high pressure engines.[61] (my italics)

Most historians do not appear ready to argue that these relationships between new and existing technologies cause innovations (appendix 3.4.3). In fact, some authors exhibit caution about asserting that a given technical factor was a necessary condition for the emergence of a given technology. Speaking of the contribution of self-acting lathes and automatic bolt cutters to screw thread production after 1860, Bruce Sinclair carefully limits his observation by saying that "improvements in machinery *made it all the easier* to manufacture a standard screw" (my italics)[62]. This is a far cry from Usher's assertion that "the complete array of prime movers *required* the further development of steam" (my italics).

References to the impact of new technology on the technological support network are less than half as common as the above-mentioned discussions of the impact of the support network on emerging technology. The thematic point is, however, the same in both cases. The historical understanding of emerging technology always stresses the interrelated character of a society's technical substructure. References to the impact of new technology on the support network fall into two classes. Some articles discuss the impact of new materials or a new process (appendix 3.4.4). Thus Dana G. Dalrymple notes the contribution of the controlled-atmosphere process of fruit storage to national marketing and distribution systems.

The lowered respiration rate [resulting from CA storage] leads to extended storage and shelf life. . . . Consequently, fruit can be held in good condition for a much longer period than is possible with conventional refrigeration alone. Formerly, for instance, preferred varieties of apples such as McIntosh could not be stored much past January; now they can be kept in fine condition until June or July.[63]

In this case the impact of CA storage is on the technological system used by fruit distributors and retailers. Other articles observe the impact of a specific new artifact (appendix 3.4.5). Louis P. Cain notes that the design adopted for sewer construction in Chicago after 1885 not only solved the problem of sewage disposal but also had beneficial side effects for the city's streets.

After the sewers were laid, earth was filled in around them, entirely covering them. The packed-down fill provided roadbeds for new, higher streets. These streets were rounded in the center with gutter apertures leading to the sewer. Such streets would stay dry and could be paved, as contrasted to the mud which had plagued the city previously.[64]

In the light of TC's usage we can identify the general characteristics of this first approach to the ambience of emerging technology. On the one hand, there appears to be consensus that technological innovation is intimately related to the network of existing technical capacities both as cause and effect. On the other hand, there is no precise definition of what is included in the concept. Given the remarkable breadth of individual technical factors that can be included under the rubric of the state of the art, and the obvious nature of the relationship itself, it is not surprising that the concept is so pervasive and yet has generated so little explicit thematic interpretation.

Technical tradition

In contrast, TC's second approach is precisely defined (appendix 3.5.1). Almost every reference to ambience as a single technical tradition explicitly or implicitly adopts the same definition. The heart of the definition has been aptly articulated by Edward Constant at the beginning of his analysis of turbojet technology; note that his term for technical tradition is "technological paradigm."

We define a technological paradigm as an accepted mode of technical operation, the usual means of accomplishing a technical task. It is the conventional system *as defined and accepted by a relevant community of technological practitioners.* A technological paradigm is not just a device or process, but, like a scientific paradigm, is also rationale, practice, procedure, method, instrumentation, and *a particular shared way of perceiving a set of technology. It is a cognition.* . . . A technological paradigm is passed on as *a tradition of practice in the preparation of aspirants to its community membership.*[65] (my italics)

Three aspects of Constant's definition are significant. First, a technical tradition is primarily a cognitive reality. It is a specific way of perceiving a technology that integrates the pertinent devices and techniques into a meaningful whole, a "governing conception." As such, the governing conception is more important to the tradition than any single element in it. Second, the governing conception is radically dependent on a community of practitioners who, recognizing themselves to be such, create the governing conception precisely because they must define the norms to be satisfied by aspirants to community membership. Third, the governing conception can be defined as the body of learning handed down from one generation to the next.

Two functions of a tradition's governing concept are obvious: it defines the meaning of all devices and techniques that are components within it, and it defines the canons of membership in the community. Upon reflection we can identify a third function by which the concept fosters the inventor's creative insight. The knowledge handed down in a tradition creates, in the minds of those who have learned it, a frame of reference with recognizable boundaries, the right and wrong ways to do the technical thing. That intellectual framework is the necessary background against which new insight emerges.

Inventive insight is always the creation of a new design concept but, in the view of TC historians, such insights always occur within some historical context that receives its focus from a tradition of praxis. The distinction between marginal and breakthrough inventions is pertinent here. A study of TC references to breakthrough inventions suggests that the authors perceive the causal nexus between tradition and invention to be no less essential for breakthrough than for incremental inventions. The historical interpretation of invention argues that inventions are not created ex nihilo. Far from being purely abstract new designs, they emerge from a richly textured historical situation, a situation that holds design constraints in tension with the human fabric of a community of practitioners.

Eighteen articles describe the cognitive function of tradition as it operates in incremental inventions (appendix 3.5.2). For example, Arthur H. Frazier's study of the cup-type river current meter includes the following observation: "When using the Ellis meter for measuring the flood waters that year [1882] Price conceived a method whereby silt-laden water could be prevented from entering its bearings."[66] William Gunn Price's invention is incremental; it depends upon and modifies the cup-type meter originated by Daniel Farrand Henry and modified by Theodore G. Ellis. Price "conceived" the new method while using the Ellis meter. He was, in other words, part of the community of practitioners created by the new device, and his intimate knowledge of the design of the meter was the frame of reference in which his insight occurred.

Twelve articles treat the relationship between a technical tradition and breakthrough inventions (appendix 3.5.3). The following quotation from Milton Kerker's study of the Newcomen steam engine is remarkable in that it includes an analysis of the role of tradition for a single breakthrough invention as well as for subsequent incremental inventions.

And so the stage was set for the great act of invention by Thomas Newcomen, using the principles first arrived at by Papin. Professor Usher tells us that "the invention of the atmospheric engine in this form was the greatest single act of *synthesis* in the history of the steam engine and must be regarded as one of the *primary or strategic inventions. The important work of Watt and his contemporaries was *critical rather than synthetic*, and though new devices had to be invented, they are, after all, *improvements of the Newcomen machine. . . .*"

We have italicized Professor Usher's use of the word synthesis in order to emphasize that there were *at least two lines of development that culminated in Newcomen's work*. On the one hand, there was the line from the air pump to the piston-in-cylinder principle in the hands of von Guericke, Hooke, Boyle, Huyghens, and Papin. But there was also unceasing effort throughout the seventeenth century to build a practical engine utilizing della Porta's concepts directly; here we need only refer to the high points associated with the names of Salomon de Caus (1576–1626), Edward Somerset (1601–1667), and Thomas Savery (1650–1715).[67] (my italics with the exception of the word "synthesis")

Following Usher, Kerker understands Newcomen's design insight to be a breakthrough invention, a "primary or strategic" insight that created a new technical tradition within which later inventors operated as incremental inventors. Usher and Kerker recognize the critical role of previous technical traditions for the breakthrough. They describe Newcomen's insight as "an act of synthesis," thus calling attention to the fact that Newcomen could not have created the new design concept unless he had intimate knowledge of the two traditions mentioned at the end of the passage. That knowledge was Newcomen's starting point. His invention broke from both separate traditions by integrating them into a single new tradition whose governing conception was his engine design.

In other studies of inventive behavior particular attention is paid to the mechanisms by which technical knowledge is handed down within a tradition (appendix 3.5.4). Robert P. Multhauf, in his treatment of the French crash program for saltpeter production, notes the dual role of a prize contest sponsored by the Académie Royale des Sciences in 1778. The contest focused inventive attention on saltpeter production and, in the process, served as a societal structure for gathering together and transmitting the knowledge that was already part of the tradition.[68] Lynwood Bryant describes the role of newspapers, scientific and technical journals, and meetings of learned societies for transmitting information in the new field of thermodynamics after 1850 in Europe and the United States.[69]

Several authors imply the critical role of tradition (i.e., as handing down knowledge) by noting the consequences of a breakdown in the process (appendix 3.5.5). Dana G. Dalrymple traces a series of experiments on controlled-atmosphere storage of fruit back to 1820 and asks why a century of work yielded so few results.

> In retrospect, the critical problem with these early observations was that they were independent events. *No one added them up.* They did not, with one exception, lead to any further work on CA storage; in fact, they remained essentially *unknown to those who followed.*[70] (my italics)

Although the CA experiments were all focused on the same process, they did not constitute a technical tradition because of their isolation from one another. They would not lead to significant invention until they became a tradition in the twentieth century.

One article, nearly unique in TC, suggests that tradition, far from being essential in the creative process, actually hinders invention (appendix 3.5.6).

> Any practice that adheres too strongly to tradition is sure to be conservative and in this way to stand in the way of improvement. Thus tradition, which once was held to be the one indispensable ingredient in any article made by man, now is regarded merely as a primitive skill which may stand in the way of a more efficient one.[71]

Many TC historians would agree with the observation that too strong an adherence to tradition can block an inventor from making breakthrough inventions.[72] What is remarkable here is the absence of any reference to the positive role of tradition in creativity. As we have seen, TC's historians tend to stress the positive rather than the negative role.

A group of articles in the internalist tradition approaches the relationship between tradition and invention in less thematic fashion. Ten articles, each focused by the classical internalist task of verifying the true origins of a specific invention, follow a series of inventions that embody a particular tradition as it unfolds over time (appendix 3.5.7). Thus Silvio A. Bedini studies a series of clepsydrae from 1277 through 1824, carefully assessing the evidence establishing the influence of the prior tradition on each new clock.[73] These articles confirm the cognitive role of tradition in inventive behavior without explicitly refering to it. Of less thematic interest are eleven articles describing a sequence of events in a technical tradition without any analysis of how prior events influenced those that followed (appendix 3.5.8).

When we turn from tradition-as-prior-to-invention references to invention-as-prior-to-tradition references we see a pattern of considerable consequence. Articles treating the subsequent influence of inventions are of two types. A few references to breakthrough inventions explicitly indicate that the invention in question creates a subsequent tradition, for instance, Milton Kerker's discussion of the Newcomen engine (appendix 3.5.9).[74] (The majority of breakthrough invention studies, however, are less concerned with the impact of the breakthrough invention than with the forces leading to it. Lynwood Bryant, for example, in his study of the 1876 Otto engine, is clearly much more concerned with the forces that coalesced in Otto's inventive breakthrough than with a detailed analysis of its results.[75])

The second type of reference is made in passing in articles whose central interest lies elsewhere (appendix 3.5.10). Thus Lee M. Pearson, after a detailed study of causes for the disastrous explosion of the artillery piece named "Peacemaker," notes that the tragedy fostered some later testing of gun designs and relevant metals.[76] It appears, therefore, that TC historians have a common bias toward seeing invention as the *end* of a process rather than the *beginning* of one. Even the internalist studies such as Bedini's analysis of a five-hundred-year tradition of water clocks reveals the same predisposition. As he moves from clock to clock he appears to leapfrog ahead to the next clock in the series and look back upon the factors influencing its design.

We can contrast technical tradition, which TC authors relate primarily with invention, and the technological support network, which they tend to relate with innovation. In the tension between innovation and the technological support network the central issue is a question of "fit." Does the potential innovation fit the constraints of its infrastructural ambience? These constraints are numerous and multifaceted, and it is the task of the entrepreneur to judge whether the fit is sufficent for the innovation to be viable. The constraints of infrastructure are not causes of the innovation except in the negative sense that their "cooperation" is a necessary precondition for successful innovation.

The primary issue in the relationship between tradition and invention is the creative tension within the inventive mind. The cognitive constraints of tradition and the inventor's insight are seen as two essential components of the inventive act. The constraints are well organized and well known to the inventor because s/he is a participant in the traditional community. While the inventor may well consider the existing infrastructure and ask whether the new insight will fit into it, as inventor s/he focuses on new possibilities, possibilities that are directly related to the limits of the governing conception of one or several traditions as s/he has come to understand them. In contrast to technological support networks, therefore, technical traditions tend to be understood as direct causes of invention and not merely as necessary conditions. We turn now to the third way that members of TC's constituency have interpreted the technological ambience, to a remarkable concept that in many ways integrates the distinct elements of infrastructure and tradition.

Systems

Systems in many respects resemble machines. A machine is a little system, created to perform as well as to connect together, in reality, those different movements and effects which the artist has occasion for. A system is an imaginary machine invented to connect together in the fancy those different movements and effects which are already in reality performed. . . . It often happens, that one great connecting principle is . . . found to be sufficient to bind together all the discordant phenomena that occur.—Adam Smith[78]

This passage from a founding father of economic theory serves as a frontispiece for our discussion of systems. Adam Smith's use of the machine—which "connects together" different movements into a single operation—as a metaphor for a theoretical system is an early statement of a principle adopted by many TC authors and reshaped as a major theme in their intepretation of emerging technology (appendix 3.6.1). In the first issue of the journal Peter F. Drucker articulates the same principle.

Technology, in other words, *must be considered as a system*, that is a collection of interrelated and intercommunicating units and activities. We know that we can study and understand such a system only if we have a *unifying focus* where the interaction of all the forces and factors within the system *registers some discernible effect*, and where in turn *the complexities of the system can be resolved in one theoretical model.*[79] (my italics)

Drucker stresses one of the two major dimensions of the system's approach as we find it in TC. The heart of a system is a "theoretical model," "a unifying focus" that serves as the principle of integration for all of its components. To understand a system's theoretical model is to achieve a vantage point from which the interaction of system components becomes intelligible; they become "visible" as elements in a single functional artifact.

Does the systemic theoretical model *perceive* functional integration or *cause* it? Is it possible to bring order to a disorganized technical situation simply by thinking of it under a single functional rubric? The answer of the TC historians who adopt a systems approach appears to be "no," for they almost invariably combine the first characteristic of systems—the governing theoretical model—with a second. By its nature systemic thinking tends toward interventions aimed at bringing previously disparate factors into functional conformity or at redesigning components to improve their conformity. Put another way, a systems approach to disorganization is a powerful type of emerging technology. The governing theoretical model not only allows the observer to "see" components as interactive, it also permits him or her to identify critical problems that must be solved for the situation to function as a single system. Thomas P. Hughes has focused particular attention on this process by adapting a military metaphor. "The evolving technology of the high-voltage transmission system . . . can best be explained by reference to needs arising *within that system*. . . . These needs can be designated as *reverse salients* in an expanding technological front" (my italics).[80] In an advancing battle line a "salient" is a bulge that occurs when one segment of the army moves forward more rapidly than the segments on either side. A "reverse salient" is a similar bulge that occurs when a segment of the line lags behind. When a single systemic model allows the observer to perceive a set of technological changes as a single "advancing front," it is possible to identify critical lag points demanding improvement.

This systems approach can foster technological change in a second way, which is related to Hughes's explanation without being identical with it. Edward Constant has pointed out that functional failures—Hughes's reverse salients—are one of two ways in which a system's point of view can generate new technology. The other occurs when the systemic vantage point allows the observer to see that "a radically different paradigm will do a much better job or will do something entirely novel."[81] In this case the systemic overview serves as a background against which the observant inventor can see a breakthrough invention that is not so much demanded by functional failure as it is made possible by some design limitation perceived by the inventor.

Seventy TC articles use a systems approach to emerging technology. Many others use a language implying similar thinking. In almost every instance the two aspects of the systems approach are identifiable, namely: the governing concept allows an observer to understand all potentially disparate factors as integrated components, and the same concept focuses attention on technological change that is either demanded or possible. The potential technological change that is made visible by the governing concept can be either incremental (an improvement within the existing system) or breakthrough (a new system, which may compete with the existing one).

This definition of the systems approach seems at first to be identical with the definition of "technical tradition" just discussed. In both, the central factor is a governing concept leading to incremental or breakthrough inventions. Further reflection, however, reveals a key distinction. A systems approach is not the only way to integrate disorganized activities. While the systemic approach seeks a hierarchical and controlling solution to disorder—Adam Smith speaks of "binding together all the discordant phenomena"—a technical tradition stresses negotiation in which the several independent factors in the "disorganized" situation are represented by people who see themselves as peers. Through negotiation, a group of technological practitioners can create a new technological integration without arranging themselves into the kind of hierarchical control-conformity relationship that often seems inherent in a systems approach.[82]

Thus, while the dynamic of technical traditions can be thought of an one example of the systems dynamic, to include it under the systems heading significantly distorts TC's interpretation of traditions. The governing concept in a technical tradition defines correct and incorrect behavior as the norm for admission into a *community* of practitioners. It includes the design characteristics of devices or techniques because the community is created by some shared technological endeavor passed down through time. Nevertheless, the purely technical aspects of the tradition—knowledge of design characteristics of devices together with appropriate skills and techniques—are inextricably part of the human fabric of the community. By contrast, the governing concept of a system is focused exclusively on the technical function. Thus, traditional norms governing individual praxis would be included in a systems model only insofar as the individuals are understood to be *functional* components. A traditional community would be a "system" only when its communal dimension was ignored.

Because the governing concept of a system is focused on function rather than community, and on hierarchical conformity rather than peer negotiation, it is possible to apply the systems approach to artifacts on many levels of complexity. The term "artifact," as used in this book, is defined according to its root meaning: "any integrated entity made by human beings." Thus "artifact" can mean a single machine, a developmental project, an entire factory, a communications network, or an economic strategy. TC historians have applied the approach on all these levels. As we consider examples of each, it will become clear that the degree of systemic complexity changes the content of the systems approach.

System as single machine. Articles that apply a systems approach to a single machine tend to stress how critical problems in a given design model give rise to inventive behavior (appendix 3.6.2). Lynwood Bryant studies the successful inventive concept leading to the 1876 Otto engine. While he discusses the developmental work involved in achieving the Silent Otto, he is most interested in the critical design concept itself and in Nicolaus Otto's inventive activity. He contrasts the capabilities of the steam engine and the internal combustion engine, stressing the economic advantages of the gasoline engine in small-power, intermittent-use situations. It was, however, a problem inherent in the design of the engine, the achievement of a smooth flow of power, that became the specific focus of Otto's inventive thinking. Throughout the years of labor described by Bryant, this problem led Otto to a succession of theories and design concepts that eventually provided a solution. In his interpretation Bryant adopts a two-level systemic framework. The potential marketability of the new engine explains in part Otto's motivation for attempting the invention. On this level the system includes the cost and marketing factors of the engine as an innovation that might compete with the steam engine market. Within this larger systemic context the engine design itself is a subsystem that played a key role in Otto's inventive process. His ability to see the engine as a functional whole allowed him to identify the critical problem calling for inventive solution: how to achieve a smooth flow of power.[83] This pattern of including smaller systems in larger ones is common in TC.

System as a development project. TC authors invariably think of development projects as systems. Typically, development is the work of a highly structured team led by an authoritative decision maker who defines the project's goal, the heart of development's systemic character. A project goal operates—precisely in the manner of a system's "governing concept"—to create an intelligible frame of reference for identifying problem areas for the developmental team to solve through model building and testing. Indeed, development projects appear to be the best-articulated examples of the systems approach in TC.

The earlier discussion of development noted that a project's stated goal changes repeatedly in response to problems encountered during testing. The project's governing concept is dynamic not only because it induces change, but also because it is continually transformed as a result of new data from the testing process that it governs. The systems approach fits development better than it does the inventive process. In the case of the Otto engine, for example, the ability to see the entire engine as a functional system identifies the critical problem that Otto had to solve, but it does not explain the mysterious process of inventive insight itself. In examples of development, however, the ability to see the project as a single goal-directed system not only identifies critical problems but structures and restructures the very procedure itself. In other words, a systemic approach *assists* the act of invention, but it *is* the process of development.

Given the perfect correlations between the systems approach and the concept of development, it is not surprising that studies of development projects frequently use the term "system" explicitly (appendix 3.6.3). They often identify system configurations on several levels of complexity just as Bryant's Otto engine study did. Thus Windham D. Miles notes in his treatment of the Polaris project that the various subordinate systems of the submarine had to be integrated into one overarching system before they had been developed because all components were being worked on simultaneously.[84]

System as production facility. When TC historians study full-scale production units such as textile mills or arsenals, they frequently interpret technological change by writing of the entire production unit as a single system (appendix 3.6.4). David J. Jeremy, in his study of early American textile mills, uses systems language explicitly and articulates a hypothesis in which innovation strategy operates systemically.

Commitment to mass production [by New England mills] had liberating consequences for *the matrix of innovation*. On the one hand, it reinforced the need for well-equipped workshops. On the other, it reduced the number of engineering problems that would arise in modifying British-type equipment because large-scale production combined with product specialization considerably limited the number of operations any given machine would be required to perform.[85] (my italics)

Jeremy argues here that a strategic decision of the New England manufacturers—to compete with British imports in a few areas of mass production—directly affected the design of the factory and the machines in it. He identifies a series of labor-saving innovations that "almost add up to a *system* of modern mass production" (my italics).[86] More important still, he describes critical innovations that integrated the various factory machines into a single process. The innovations were made piecemeal, each one the result of a new bottleneck in the production line caused by speeding up some other machine function. His description of the system dynamic at work is strikingly similar to Hughes's "reverse salient" model.

> At least two production influences *gave momentum* to these innovations springing from the interaction of American market forces and factor endowment. The first arose from the early and unavoidable reliance on empirical rather than scientific knowledge in textile manufacturing. Because productivity *of each processing step* could not be precisely predicted, *imbalances* appearing between production capacities of machines at various stages *impelled* improvements at the levels *falling behind.* (my italics)

Finally, Jeremy notes that these technical imbalances combined with economic pressures on the company to *cause* subsequent innovations.

> By 1831 when nine companies existed at Lowell, agents had become aware of competition in terms of production . . . and cost comparisons provided measures of performance which, in the mind of the . . . agent, appeared to be related to improved equipment. Making this connection, *it was natural to turn to technological improvements* as the internal answer both to production imbalances and to the interlocking directorates' pressure for rising dividends when a major capital expansion could not be justified.[87] (my italics)

Here we see three levels where the system dynamic operates. Individual machines are systems that can be improved by innovations to save labor or increase speed. The entire set of machines, when seen as a system, reveal production lag points calling for innovation. Third, on the level of the company as a profit-making enterprise, the systems dynamic operates both in terms of the strategy of high-volume production of a few targeted grades of textiles and in terms of the demand for increased output without significant capital expansion. System components need not be limited to material parts. Insofar as a company—with its array of machines, raw materials, workers, finances, management, and strategies—is understood as a single functioning whole, it fits the definition of a system as precisely as a single machine.

Jeremy does not tell the tale of the specific inventions or developments in the Lowell mills in any great detail. His three-tiered systemic model stresses innovation strategies while including acts of invention and development by implication. As in every case thus far, he understands the systems dynamic as a set of increasingly inclusive and hierarchical systems.

System as a transmission network. A wide variety of transmission networks—rail and air transportation networks, sewage systems, trolley lines, electric light and power grids, power transmission systems in factories, telephone and telegraph networks—have been studied by TC authors (appendix 3.6.5). The definition of transmission networks, as we find them developing within industrial capitalism, has tended to stress the requirement of systemic conformity if the network is to operate effectively.[88] A national rail system, for example, cannot function without standardized track gauge and car-coupling designs, without adequate managerial and accounting procedures, and without a system-wide information network to provide management with data about the location and speed of trains and the contents of their cargo. TC's discussions of transmission networks tend to be less explicit in their use of system language than are case studies of development or production systems. They can be considered part of the systems approach, however, because their interpretation of emerging technology implicitly embodies one of the major characteristics of the systems approach, that is, functional problems within a system calling for specific innovative solutions.

Charles H. Clark provides a typical instance. He recounts the twenty-year struggle of the rail industry to foster the invention and development of an acceptable freight car coupling unit that would reduce time loss and equipment damage as well as answering the call for worker safety arising from numerous ugly accidents. The rail companies' problem was complicated after 1865 by their desire to agree on a single standardized design to permit interchangeable freight cars on different lines.[89] Clark notes that the entire process took place at the same time that the railroads were working out other problems of standardization, such as car design and track gauge, which the increasing integration of the several lines demanded. Thus, while the malfunction of the car-coupling unit (a single mechanical unit seen as a system in itself) focused attention on a need for invention and development, the railroads' commitment to an integrated national network focused their attention on a requirement of production units and transmission networks alike, standardization of system components (appendix 3.6.6).

This requirement is often noted by TC authors. If a large-scale and complex set of technologies is to function as a system, physical standardization of all components tends to solve the problem of bottlenecks, those gaps between interconnecting but nonstandardized technologies that must otherwise be bridged by manual labor operations that are, from a systems point of view, costly and time-consuming.[90] In Clark's case study the cooperative search by many rail companies for a standardized coupler flows from a "governing concept," their perception of the national rail network as one functional unit wherein all cars must be interchangeable. As in our earlier examples, Clark portrays systems dynamics at work on several inclusive levels.

System as a sector of the economy. We have been tracing the systems approach through examples of increasing levels of complexity. The final type of article in this series approaches an entire segment of a national or regional economy with language suggesting that the economy is a system. Six TC articles refer to the agricultural sector, four discuss the manufacturing sector, and two study state-planned economies (appendix 3.6.7). As with discussions of transmission networks, these articles do not ordinarily use explicit systems language. Nonetheless, the congruence between their language and the systems model is an indication of the potential range of the concept for interpreting emerging technology.

In all three types of study the economic system in question includes an array of inventors, technical traditions, development projects, transmission networks, and production units. The set of subsystems is so broad that it taxes the notion of system as we have seen it thus far. In all previous examples the role of the governing concept (i.e., as inducing technical change) has been a focal point for systems interpretation. In studies of entire economic sectors, however, the integrating function of the single governing concept tends to be assumed rather than explicitly discussed. The two articles about Russia's planned economy are exceptions. They stress the role of the state's overarching economic strategy and discuss its conceptual content.[91]

The other articles assume that the role of a governing concept is played by market forces and they focus attention on them to explain innovative activity. Fred W. Kohlmeyer and Floyd L. Herum's study of science and engineering in United States agriculture is a case in point.

Although one can point to a *system* of prior inventions and slow piecemeal improvement of existing devices, it was not until the 1830's that the pace of mechanization in agriculture definitely accelerated. *The growing urban-industrial sector served agriculture* in many ways *supplying* manufacturing facilities, services, and a market for the increased output of farm commodities.[92] (my italics)

Kohlmeyer and Herum assume that the agricultural and urbanindustrial sectors are interactive but they do not articulate a market theory on which the assumption is based. The dynamics that "accelerate" the pace of agricultural mechanization appear to be typical systemic imbalances. The three factors—"urban growth," "farm output," and "innovation in farm machines"—are spoken of as if they were functional parts of one larger system. Just as a change in the speed of a new loom in a textile mill creates bottlenecks in other parts of the mill and calls for further innovation, so here change in one area of the economy (e.g., "urban growth") fosters further changes such as increased farm output and farm machinery innovations. Expressions such as "the *impetus* to mechanize" and "the *rate* of technical *progress* in agriculture" enhance the suggestion that the agricultural sector operates within the larger economy with systemic dynamism (my italics). The historical details implicit on the various levels of the system are not discussed. Thus the application of a systems approach to an entire economic sector tends to blur the details of the "subsystems" where the concrete activities of emerging technology actually reside. As a result, the unique complexities of the several types of activity involved in emerging technology (invention, development, and innovation, together with technological support networks, technical traditions, and system dynamics) are homogenized so that they can serve as quantifiable data supporting the macroeconomic interpretation being articulated.

The application of a systems mentality to macroeconomic interpretations of technological change leads, it would seem, to a replication in theoretical interpretation of the standardization process noted in U.S. network designs. As the U.S. option for standardized systems forced the redesign of components from many diverse designs into uniformity, so the option for a systems interpretation of the economy forces the historically diverse dimensions of individual moments in emerging technologies into a form of quantifiable uniformity.

This standardization and redefinition of unique historical factors can be observed in Nathan Rosenberg's analysis of economic development and technology transfer. Writing of the problem of underdeveloped countries becoming efficient producers of capital goods appropriate for their needs, he discusses the process of technical improvement within the manufacturing sector of an economy.

> Technological change is . . . a continuous stream of innumerable minor adjustments, modifications, and adaptations by skilled personnel, and the technical vitality *of an economy* employing a machine technology is critically affected by *its* capacity *to make these adaptations*. The necessary skills, in the past, were developed and diffused in large measure by the capital goods sector. The skills are, inevitably, embodied in the human agent and not in the machine, and unless these skills are somehow made available, the prospects for the viability of a machine technology may not be very good.[93] (my italics)

By saying that "the necessary skills" on which technological change depends are "embodied in the human agent" Rosenberg acknowledges the unique human texture of all aspects of technological change. His language indicates, however, that he understands "the economy" itself as a functional reality. "It" must "make these adaptations." Because his major concern is with the development of the economy as a whole, the human factors, and the individuals who embody them, are defined as economic factors, components within the larger system. They are "the *necessary* skills" (i.e., necessary in terms of their functional contribution) and they must "*somehow* be made available" (i.e., the historical context out of which each human factor emerges is not dealt with here).

This survey of the various uses of the systems approach for interpreting emerging technology reveals several noteworthy characteristics. On the one hand, TC's systems approach appears to be intrinsically hierarchical. No matter what level of complexity the central focus of an article, every TC author considered here adopts a set of systems on several inclusive levels. The system of central interest is either understood as included within the dynamic of a larger system or as including other smaller systems within itself. On the other hand, the use of a systems approach for *historical* interpretation of emerging technology appears to be most effective in the intermediate range of the levels of complexity considered here. At the level of the individual machine or invention, the model is not helpful in explaining the critically important factor of personal creativity. In addition, the functional focus of a system's dynamic downgrades the importance of communal dynamics, which we observed in the model of technical tradition. At the other end of the spectrum, on the level of macroeconomics, the use of the systems approach tends to blur the details of the historical processes at work in its many subsystems. By contrast, the approach seems particularly apt for explaining the developmental process and the innovation strategies adopted by management in manufacturing firms and transmission networks.

Table 10
Correlation of methodological styles and references to ambience

	Support networks	Tradition	Systems
Contextual	44 (66%)	45 (71%)	61 (86%)
Internalist	8 (12%)	13 (21%)	3 (4%)
Externalist	5 (8%)		3 (4%)
Nonhistorical	7 (11%)	3 (5%)	3 (4%)
Historiographical	2 (3%)	2 (3%)	1 (1%)

As the quotation from Adam Smith at the head of this discussion suggests, there appears to be an inherent, although unconscious, link in TC's systems analyses between the systemic style of hierarchical control and standardization and the ideological commitment of industrial capitalism to the same values. It should not be surprising, then, that TC's systems model is incompatible with its technical tradition model or that the systems approach has proven to be most apt for discussions of cost-intensive developmental and innovation phases of emerging technology. Finally, the use of a systems approach in macroeconomic interpretations is also clearly congruent with a capitalist world view. Other aspects of the systems approach will become clear in the following discussion.[94]

Historiographical Status of the Systems Approach

Within the seventy articles adopting a systems approach for emerging technology, several historiographical characteristics are worth attention. The articles show remarkable consistency in defining the role of the systems dynamic in technological change. At the same time, the proportion of systems references found in contextual articles is striking (table 10).

References to the technological support network and to technical tradition reveal their links with the earlier historiographical tradition of internalist history and, in the case of the support network references in nonhistorical essays, with economic theory. By contrast, TC's references to a systems approach are overwhelmingly contextual, exceeding even the high percentages found in references to invention (68 percent), development (83 percent), and innovation (81 percent). We have already noted the thematic creativity of the contextual historians in our discussion of the tripartite model of invention–development–innovation. The remarkably high contextual share of systems references strongly suggests that TC's contextual historians have not simply borrowed the systems approach from an earlier economic tradition, but have reshaped its meaning and appropriated it as a historical theme. We can ask, therefore, what makes the approach so attractive for historians of technology.

The systems approach is a provocative example of a parallel discussed in chapter 1 between actual technological practice and historical reflection on that practice. It is difficult to distinguish, in many TC treatments of systems, whether the governing concept that integrates the system is of greater help to the individuals active in technological change or to the historians seeking to interpret their behavior. For practitioner and historian alike the frame of reference provided by a systemic overview can serve the same purposes. It is a heuristic device allowing both to perceive many potentially disconnected elements as interactive components within a functional whole. For the technological practitioner this single point of view helps to make the decisions involved in maintaining the system, in improving it through incremental inventions, or in visualizing breakthrough inventions leading to entirely new systems. For the historian it serves to create a frame of reference for the many confusing details involved in a given historical situation.

At present the historiographical status of the systems approach in TC is that of one helpful subtheme among others for interpreting emerging technology. As a way of portraying the technological ambience it stands in sharpest contrast to the technical tradition model. Where the tradition model stresses the tension between communal and personal interactive aspects of technological creativity and technical or functional aspects, the systems model stresses only the functional dimension and tends to reduce the human interactive aspects to functional standardization. Were the systems approach to become the dominant model for historians, it seems clear that its inherent presuppositions could radically transform the nature of the history of technology.

We have repeatedly pointed out that the fundamental tension between "design" and "ambience" is central to the contextual history of technology. It is the creative wellspring of new technology and, for the historian, of a richly textured interpretation of technology's historical fabric. In this chapter we have seen how the design-ambient tension operates in inventive, developmental, and innovative activity. Even more striking is the way the technical tradition model embodies the tension. The tendency of a systems approach, were it used to reinterpret that activity, would be to ignore the unique historicity of ambient factors by redefining them as functional components within a more inclusive system.

Acts of invention can be interpreted as components within a development project. It is possible to find language in TC that speaks of the "industry" of "systematic invention."[95] Again, the highly personal struggle in a development project, between a team of individuals and the recalcitrancies of the specific situation, can be reduced to a time-cost analysis where the entire project is a component within a corporation's innovation strategy. In like manner, workers within a corporation can be reduced from active human participants in the technological process to functional cogs in the standardized systemic machine. The human tensions involved in managerial choices on the level of innovation strategy can be ignored when innovation activities are defined as components in a macroeconomic system.

Thus an exclusive use of the hierarchical systems approach could well dissolve the tension between design and ambience by defining all ambient factors as functional components. Were this to happen, the historical ambience wherein reside personal creativity, aesthetics, politics, and religious faith—in short, all of the passionate, truculent, and unsystemic dimensions of human life and culture—would no longer be visible to the historian. And if we understand the act of historical research as the encounter between historian and the full range of human factors in their unique historicity, then such an assumption could be said to destroy the historical endeavor itself. Instead of helping society at large to understand its past as a human endeavor, the historian of technology would foster the myth that technological change results from exclusively functional factors governed by an autonomous technological dynamic. In such an extreme use of the systems approach, the history of technology would be reduced to a detailed chronicle of the progress of "the system" and all nonsystemic human factors would vanish. Philosopher Joseph Agassi stresses the cognitive danger implicit in a complete systemic reductionism in the following passage. Note that he uses the terms "algorism" and "system" synonymously.

> Any partial systematization, such as Descartes' [geometry] and Wald's [system of decision making procedures], merely covers some ground and sends the adventurer further afield in search of new frontiers. *But total systematization excludes all adventure.* In any field, any algorism is welcome; yet, were an algorism *universal* in that field, *creative thinking in that field would be redundant* altogether, once and for all. If the field in question is *the generation of ideas in general*, algorism in it is *the end of creative thinking in all science and technology.*[96] (my italics)

The danger of a systemic reductionism for historians of technology tends to be overstated by the structure of this chapter. Because of the complexity of the six subthemes seen here I have deliberately postponed a treatment of nontechnological dimensions of the technological ambience so that these themes could be presented with a minimum of confusion. In fact, the articles considered in this chapter, and contextual articles in particular, do not restrict their interpretations of emerging technology to technological factors. They, together with other articles not treating emerging technology, are filled with references to the nontechnological dimensions of history that constitute such a great part of the human ambience in which technology exists. Thus, although the bias toward a purely functional reductionism inherent in the systems approach bears careful consideration by historians of technology, current usage does not necessarily imply a systemic reductionism by TC authors. The complex question of how well TC authors succeed in escaping the reductionist trap is central to this study and will be addressed in detail in chapter 5.[97]

The themes to be discussed in the next two chapters, the relationships between technology and science and between technology and culture, are intimately related to the theme of emerging technology. They will provide a more detailed understanding of TC's full interpretation of the technological ambience.

3

Science, Technology, and the Characteristics of Technological Knowledge

No theme, in SHOT's early years seemed as well focused, as important, or as interesting as the relationship between science and technology. The society had, after all, been born "of a rib from the side" of the History of Science Society and most members took it for granted that the task of differentiating the history of technology from its parent discipline was not only a matter of finding new group identity but one of the critical issues for the field as well. As early as Fall 1961, TC published a nine-article "special issue" on the relationship of science and engineering, a promising beginning to be sure.

Fifteen volumes and many Science–Technology articles later, the journal took up the matter with another burst of intensity when it published the proceedings of the 1973 Burndy Library Conference on "The Interaction of Science and Technology in the Industrial Age."[1] Nevertheless, despite brave beginnings and TC's remarkable sustained interest in the topic the theme itself clearly troubled many Burndy participants. Nathan Reingold and Arthur Molella suggest, in their introduction, that "perhaps historians should approach their problems [the generation of technological novelties] with a different set of categories." Otto Mayr argues that "practical usable criterion for making sharp and neat distinctions between science and technology simply do not exist." In Eric Robinson's words, "It becomes impossible to have a useful discussion unless we have words which have significance to us now and also utility in discussing the past." Most striking of all, Arnold Thackray makes what appears to be a declaration of bankruptcy for the theme.

. . . *the categories of analysis, "science" and "technology," are not illuminating categories* for understanding these activities. If you want to talk about

Table 11
Correlation of methodological styles and references to science-technology and emerging technology

	Emerging technology	Science-technology
Contextual	287 (76%)	63 (45%)
Internalist	41 (11%)	17 (12%)
Externalist	18 (5%)	23 (15%)
Nonhistorical	19 (5%)	26 (18%)
Historiographical	13 (3%)	12 (9%)

the 18th century salt industry, you must use different terms to get at anything meaningful. That leaves me wondering how far the whole field is in a position where *the old historiographic model*, which has told us what to do and how to do it, *has broken down*.[2] (my italics)

When we consider TC's science–technology articles as a set, the pessimism of the Burndy participants appears to be well warranted. Despite the popularity of the theme—141 articles address the issue—the language of the articles reveals the problematic and confusing intellectual status of the theme for their authors. A contrast between the science–technology theme and the emerging technology theme is illuminating.

We have seen in chapters 1 and 2 that the contextual approach to the history of technology is TC's dominant methodological style not only because 50 percent of all articles adopt it, but because these articles reveal by far the greatest evidence of thematic creativity in the journal. Table 11 indicates that, while the contextual style is dramatically overrepresented for emerging technology references (76 percent), it is significantly underrepresented for science–technology references (45 percent) when both are compared with the contextual style's 50 percent of all articles. In sharp contrast, we find that TC's nonhistorical essays, clearly the most marginal methodological style in the journal, hold a disproportionately large share of science–technology references (18 percent), while they are dramatically underrepresented in discourse about emerging technology (5 percent). These contrasts suggest that TC's science–technology discourse holds a more marginal position in the journal's overall universe of discourse than does the emerging technology theme.

When we look for signs of emerging consensus we find a similar disparity between the two themes. As we saw in chapter 2, five of the six subcategories in emerging technology—invention, development, innovation, technical traditions, and systems—show a high degree of consensus by the end of the twenty-one years considered in this study. On the other hand, we find discussions of science and technology plagued by a welter of conflicting definitions of its two basic terms. Theories about their relationship abound, overlapping and contradicting one another from the beginning of the journal's publication until recent times. In short, TC historians have not developed a thematic language about science and technology that they find appropriate for the history of technology. Instead we find evidence of the disintegration of the theme despite the fact that it addresses issues that are clearly of great interest in the field.

My analysis of TC's science–technology language is an attempt to interpret this seeming paradox. The results can be summarized in advance by stating that TC's thematic confusion derives from the inappropriate frame of reference created by the science–technology question itself. Embedded in the discussion we find evidence that historians of technology are less concerned with the science–technology relationship than with the nature of Technological Knowledge. The critical evidence supporting this conclusion is found in the controversy over the formula equating technology with "applied science." Consequently, we will devote considerable attention to the sharply antagonistic positions surrounding this issue. The controversy not only clarifies the status of the science–technology theme in TC but also reveals a radically different approach to technological knowledge, which has quietly emerged from TC's historical studies and which satisfies the methodological canons of TC more than the older question of science and technology.[3]

TC's science–technology talk is extraordinarily complex. The seven statements discussed in the next section have been crafted as an attempt to include all of TC's rather disparate references to the topic. The first two contrast science and technology. The next three discuss interactions between science and technology. And the last two are simply cover terms for the opposing positions in the "applied science" controversy. The applied science controversy will serve both as a summation of TC's science–technology discourse and as an introduction to the potential new theme: Characteristics of Technological Knowledge. This theme will then be discussed in terms of four characteristics found in many TC articles—scientific concepts, problematic data, engineering theory, and technical skill. Characteristics of Technological Knowledge has not been acknowledged as a theme by TC historians at this stage of their emerging shared discourse. The four characteristics are treated as a single theme because they are all explicit or implicit critiques of the applied science hypothesis. At present, then, the new theme is only a potential candidate for the status of a major theme in the journal.[4]

The Relationship between Science and Technology

Journal references grouped under the first five statements—those describing contrasts and interactions between science and technology—are best understood in the light of the applied science controversy of statements six and seven. Some of the positions pertaining to the first five statements tend to support the applied science position while others contradict it. It will be helpful to analyze these five statements not only to clarify the many disparate observations TC authors have made about science and technology, but also as background for the applied science debate.

Statement 1: Scientific activity is motivated by curiosity, whereas technology is motivated by the desire to solve problems.

The contrast between curiosity and practicality is used by several authors to help distinguish between science and technology. Thus Richard G. Hewlett explains the tension between Manhattan Project director Leslie Groves and scientists at work under his direction by noting Groves's hidden motivation for compartmentalizing the work of each scientist, a policy ostensibly ordered for security reasons.

> He [Groves] understood that *curiosity* was an important motivation for scientific research. Channeled in the right direction, curiosity could be an asset to the project, but not if it led scientists away from their assignments. . . . Scientists in the Manhattan Project were particularly susceptible to such diversions, because much of the work in the wartime effort was dull and not essentially creative. The scientists were not *free to follow a line of investigation wherever it might lead them.*[5] (my italics)

Hewlett interprets the Manhattan Project as an engineering endeavor that needed scientific expertise, but not the uncontrolled explorations of scientific curiosity. The pragmatic demands of the task at hand were in tension with what Hewlett—and Groves—saw as the natural bent of the scientist.

While Hewlett's observation is a limited conclusion drawn from a single case study (appendix 3.7.1), we find other authors who advance the curiosity–pragmatism distinction as a universal hypothesis (appendix 3.7.2) It is normally embodied in a brief formula, stated more or less in passing. Mario Bunge includes the following statement as a working assumption for his major argument.

The method and the theories of science can be applied either to increasing our knowledge of the external and the internal reality or to enhancing our welfare and power. If the goal is purely cognitive, pure science is obtained; if primarily practical, applied science.[6]

Thomas P. Hughes and Cyril S. Smith critique the distinction insofar as it is seen as a general law covering all cases (appendix 3.7.3). Smith does so implicitly by noting instances in metallurgical history where the metalworkers' intimacy with metals is combined with curiosity about their behavior, a situation leading to new information about metals but not necessarily to scientific theory.[7] Hughes is more explicit when he points out that electrical engineers Harris J. Ryan and F. W. Peek, Jr., exhibit a blend of the two kinds of motive.

Their *enthusiasm* for their experiments is *the kind one associates with pure science*, and their experimental style is that usually associated with the scientist. Obviously, Peek worked on lightning because transmission lines needed protective devices. But it is also clear from the *enthusiastic style* of his writing that lightning had "got him."[8] (my italics)

Hughes's comment illustrates the status of this hypothesis in current historical discourse. He acknowledges the common association of curiosity with science rather than technology and, while he does not argue that the distinction is never operative, he demonstrates that historians of technology can identify instances where it breaks down.

Statement 2: The "desired artifact" in science is a theoretical model, whereas knowledge is in the service of the "desired artifact" of technology.

The second contrast between science and technology is more common than the first; it is also more complex because it treats motives as part of a larger explanatory model. Authors adopting this view normally assume that both science and technology are aimed at the creation of artifacts (appendix 3.8.1). While technology creates, modifies, or operates tools or systems, science articulates theoretical models that are as much artifacts as tools or systems. Joseph Agassi expresses the principle, citing with approval one aspect of Karl Popper's philosophy of science.

Only Popper, to my knowledge, has stressed that the ability to predict with some degree of precision is what characterizes a scientific theory—regardless of the correctness of the prediction. . . . The goodness of the hypothesis, according to Popper, may be measured by its explanatory power, as well as by its testability, which is the ability to deduce from it predictions which may be checked by experiment and observation, quite independently of whether the predictions are later found to be true or false.[9]

The object of scientific activity, in the strict sense, is not "knowledge" or "truth" but "theory." Notice that the artifact in question is constituted by the theoretical hypothesis *and* by the data considered relevant for testing its predictive ability. The truth or falsity of the hypothesis and the data supporting it are less important than the congruence between data and hypothesis. Agassi continues this line of argument by noting that the refutation of a given theoretical model does not cause the model to vanish from the resulting scientific domain. Just as old technological artifacts do not "go away" but remain as influential constraints on the new technological landscape, so too the constructions of science continue to exert influence.

The task after the refutation of the new hypothesis is not the same as before: we now wish to explain, not the same *set of facts* which the refuted hypothesis has explained, or the same *set of facts plus the refuting facts*, but *the refuted hypothesis*, as a special case and as a first approximation, plus the new facts. What Einstein explained is not all the known astronomical facts but Newtonian astronomy plus the facts it failed to explain. . . . This is a fundamental methodological point. The only way to evade it is to claim that *pure empirical data exist*; as you may know, such claims are becoming increasingly difficult to maintain.[10] (my italics)

Scientists, according to Agassi, do not find pure facts and then explain them theoretically. They construct theories that define and even design facts to fit the theory.

Along with philosophers such as Agassi, a number of historians of technology contrast science and technology in a similar fashion. Cyril S. Smith describes the nineteenth-century contrast between metallurgists and chemists as follows:

The nineteenth century illustrates well the fundamental difference between the attitude of the technologist and of the scientist. *The technologist is concerned with complex materials in their actuality*, and his knowledge will usually be empirically gathered, for it is likely to involve far more factors than can be treated exactly by the scientist. . . . Berthollet had rightly insisted that many compounds were *not* [author's italics] of fixed composition, *but the advance of chemical theory required blindness to things which did not fit*.[11] (my italics)

Smith contrasts the metallurgist, concerned with all the variables impinging on the metalworking process, and the chemist, concerned with the advancement of a specific theory and with only those variables helpful to the theory. Implicit in his observation is Agassi's thesis that science is best understood as the construction of an artifact, that is, a theory and congruent data. In Smith's example, however, the contrast between science and technology is more sharply drawn than in Agassi. The metallurgist exemplifies the technological focus on a practical and functioning artifact as the final desideratum. The chemist exemplifies the scientific focus on a theoretical artifact as the desired goal.

This contrast is highlighted by Edwin Layton when he compares the role of theory for scientists and engineers. Each group of professionals adopts a theoretical approach to the study of strengths of materials, but the theories differ radically because their purposes are different.

> Scientists tended to explain their findings by reference to the most fundamental entities, such as atoms, ether, and forces. But these entities cannot always be observed directly. To be useful to a designer, however, a formulation must deal with measurable entities, particularly those of importance to the practical man. These need not be fundamental in the scientific sense. . . . Scientific theoretical models were also needless complications from the technological point of view. . . . Engineers were content with a simple macroscopic model—for example, viewing a beam as a bundle of fibers.[12]

Smith's and Layton's comments are complementary. Smith suggests that the set of variables in metallurgy is too complex for the chemist's development of a given theory, while Layton argues that the scientific theory of strengths of materials is too complex for the engineer's purposes. These contrasting complexities underscore the basic premise of statement 2: the conceptual structure of the data used in science and in technology is governed by the very different artifacts that are the desired goal in each endeavor.

We should note in conclusion that there is no consensus among authors adopting this hypothesis about a distinction—which will be clarified later in the chapter—between theoretical engineering knowledge and nontheoretical or empirical knowledge (appendixes 3.8.2 and 3.8.3). Smith writes of "empirical knowledge" while Layton writes of "engineering theory." This ambiquity is inherent in a thematic model whose primary intent is to distinguish between science and technology. The overriding importance given to that relationship tends to blur significant distinctions between types of technological knowledge. This is one of many indications of the limitations of the Science–Technology question for historians of technology.

Statement 3: Science fosters technological creativity and rationalizes existing technological practice.

The first two statements identify the most common ways in which TC authors contrast science and technology. The next three statements summarize their discussions of how the two interact. Statement three is a cover term for references to the positive contributions of science to technology. They can be divided into hypotheses based on individual case studies and broad generalizations.

Several case studies of inventions point out that a scientific theory was a necessary precondition for a given insight (appendix 3.9.1). Thus the invention of the loading coil for long-distance telephone systems depended on an understanding of Maxwell's theory. Here scientific knowledge constitutes part of an intellectual frame of reference, a context within which the inventive insight emerges.[13]

Other articles describe the scientific contribution to emerging technology in developmental projects (appendix 3.9.2) These authors see the developmental process as governed in part by the scientific knowledge or methodology of those working on the project. Sometimes science helps to critique a set of inventive designs that have emerged after a scientific breakthrough. Lynwood Bryant sees scientists serving as critics of the inventions that appeared after the creation of thermodynamic theory in the nineteenth century.

> Some pioneers in [thermodynamic] theory such as Rankine and Redtenbacher invented engines, but the literature shows the professors more often *testing and evaluating the engines designed by the practical men.* One can see them assuming the role of consultants to successful engineering enterprises. They seem to be critics rather than creators.[14] (my italics)

More commonly, TC authors stress the scientific component of work in actual developmental projects. Otto Mayr, for example, indicates that Charles Porter's original insight into the inventive design of the high-speed steam engine was not due to his scientific knowledge, but that this knowledge was important for guiding the development of the design concept into a successful engine.[15]

Other authors discuss cases where scientific theory creates a theoretical rationalization for some older technological practice (appendix 3.9.3). Robert P. Multhauf describes Lavoisier's contribution to an understanding of the chemistry involved in saltpeter production.

One can see in the writings of French chemists, beginning with Lavoisier, a shedding of the Stahlian notions . . . and their replacement by a correct analysis of the composition of nitric acid and of the several bases . . . with which it combines. Indeed, the modern processes for the synthesis of nitric acid were known in Paris by about 1790.[16]

Multhauf is typical of other case studies of this kind in pointing out that the new scientific rationalization provided by Lavoisier and others was not sufficient by itself to generate a technical breakthrough in the actual production of saltpeter: "But important as these discoveries were *for science*, they did not contribute for another century *to the solution of the saltpeter production problem*" (my italics).[17] Production problems demanded engineering expertise.

These case studies agree on two points. On the one hand, they identify the scientific contribution to technology as the capacity to provide theoretical frames of reference for understanding technical processes. On the other hand, science is not seen to be a sufficient cause for technical innovation. It should also be noted that case studies of this sort do not attempt any generalizations about Science–Technology interactions. Their conclusions are limited to the case in question.

A number of authors, however, make general statements about technology's debt to science (appendix 3.9.4). The following passage from A. Rupert Hall is representative.

The late eighteenth century was the point in time at which the curve of diminishing returns from *pure empiricism* dipped to meet the curve of increasing returns from *applied science*. This point we can fix fairly exactly, and so we may be sure that if science had stopped dead with Newton, technology would have halted with Rennie, or thereabouts. The great advances of later nineteenth century technology *owe everything* to post-Newtonian science.[18] (my italics)

Hall appears to argue that the individual contributions of science to new technology seen above in case study form are examples of a universal law governing all modern technology, namely that all technological innovation depends on prior science. His hypothesis and others like it are examples of the applied science hypothesis and will be discussed in detail below. They are mentioned here to indicate that they are a common result of attempts to generalize from single cases about the impact of science on technology.

Statement 4: Technology contributes to science by creating instruments, by posing scientific problems, and by creating new conceptual models for later science.

Discussions of technological contributions to science, which complement the hypotheses of statement three, can be summarized under the three divisions of statement four. The language commonly used for the first two suggests that TC authors take them for granted as commonplace technological contributions to science. Technical traditions of craftsmanship and mechanical invention are seen as the origin of instruments critical for the scientific revolution in Europe (appendix 3.10.1). Silvio A. Bedini introduces a body of evidence substantiating the contribution of automata to scientific instruments in the following terms: "It is relatively simple to trace the evolution from the craft of the clockmaker to the art of making fine instruments, which became a dominating force in the Scientific Revolution."[19] Here the technology–science link is considered obvious; Bedini's major concern is to verify the causal connection in the case of automata, not to demonstrate the general pattern. Other authors make a similar point about later technological contributions. Hugh L. Dryden notes in passing that space satellites are actually scientific instruments for conducting experiments in space.[20]

The same common acceptance can be seen when authors note that technology poses new problems for science (appendix 3.10.2). Some of these references take the form of assertions accompanied by examples. John B. Rae's statement is typical.

Nor does it follow that the pragmatic approach necessarily precludes research in fundamentals. Kendall Birr's scholarly study of the General Electric Research Laboratory makes it clear that a good deal of "basic" research grew out of efforts to find a solution to a specific problem.[21]

Lynwood Bryant observes that the process of combustion in gasoline and diesel engines continues to pose theoretical problems for science well after the adoption of the engines themselves.[22]

Several authors discuss a technological contribution to science that is radically different from the two commonplace observations just cited (appendix 3.10.3). In his study of the influence of the eighteenth- and nineteenth-century traditions of steam and water engines on Sadi Carnot's thermodynamics, Donald Cardwell stresses the conceptual congruence between machines in the two traditions and in Carnot's model.

The source of the *concepts* [Carnot's] was unquestionably technological, and this constitutes a very serious objection to *the theory that the de-*

velopment of "pure" science owes little or nothing to technology. Also, it might be argued that, unless there is some account of the technological factors, the *concepts* of thermodynamics, as described in standard works on heat, must seem curiously arbitrary inventions.[23] (my italics)

Cardwell sees the conceptual link between the physical design of engines and the theoretical structure of Carnot's model as a falsifying example for the hypothesis that pure science is never *conceptually* indebted to technology. Scott Buchanan makes the same point while discussing the Greek insight that art imitates nature. He advances a generalization with implications for contemporary science.

Art or technology thus becomes the midwife of science. If you want to understand something, make a similar object or artifact; then impute that artistic process of making to nature. *We are not far from this in our current use of models in science.*[24] (my italics)

This conceptual contribution to scientific theory is remarkable because it places technology in the role of helping to shape or create scientific theory much as scientific theory sometimes plays a creative role in new inventions. This is, of course, a much more substantive contribution to science than merely serving as instrument maker or problematic object of study providing grist for the scientific mill.

Statement 5: Scientific and technological activities that occur in human communities often influence science-technology interactions.

Some TC authors explain the interactions of science and technology by studying the historical communities in which these activities have taken place. Several of them see the distinction between scientific and technological communities as relatively clear-cut (appendix 3.11.1). Edwin Layton has argued that the communities took on "mirror-image" characteristics in late nineteenth-century America.

While the two communities shared many of the same values, they reversed their rank order. In the physical sciences the highest prestige went to the most abstract and general—that is to the mathematical theorists from Newton to Einstein. Instrumentation and applications generally ranked lowest. In the technological community the successful designer or builder ranked highest, the "mere" theorist the lowest. These differences are inherent in the ends pursued by the two communities: scientists seek to know, technologists to do. These values influence not only the status of occupational specialists, but the nature of the work done and the "language" in which that work is expressed. . . . For information to pass from one community to the other often involves extensive reformulation and an act of creative insight. This requires men who are in some sense members of both communities.[25]

Several points are noteworthy. The two communities are independent sources of theory, each with its own "language," and they enter into mutual sharing through the mediation of people fluent in both. The distinctive character of the two languages comes from the reversal of priority between abstract theory and the ability to design technical artifacts. Thus, for Layton and others adopting this approach, the interactions and contrasts between science and technology result directly from the value systems of the historical communities that have become the institutional loci for the two minds of activity.

Other authors express more hesitation about the historian's ability to define the boundaries between scientific and technological communities (appendix 3.11.2). At the heart of the problem lies the fact that individuals have defined the two terms differently at different times. In his study of pre-nineteenth-century salt production, Robert P. Multhauf underscores the confusion of language that frustrates attempts to label some actors "scientists" and others "technologists."

As one proceeds in a study such as this he sees increasingly clearly what individuals were doing, or thought they were doing, *while becoming increasingly bewildered as to what is science and what is technology*. We may prefer to talk about scientists and technologists. But whereas rather too many people will refer to themselves as scientists, no one ever seems to have called himself a technologist. He has given himself names—inventor, mechanic, engineer—but their very instability certifies their inadequacy to characterize the technologist.[26] (my italics)

The fact that scientific–technological activities occur in definable communities does not necessarily mean that the communities or the activities can be neatly divided by the terms "science" and "technology." In a similar vein Nathan Reingold and Richard A. Overfield speak of the "gray area" and the "blurred distinctions" between science and technology in Alexander Dallas Bache's Coast Survey and Charles E. Bessey's group of botanists at the University of Nebraska.[27]

The linguistic confusion of the science vs. technology theme

In an article devoted to the historiographic problems inherent in the Science–Technology distinction, Otto Mayr aptly summarizes the resulting confusion, not only in the attempt to label specific historical communities as one or the other, but in all of the hypotheses, based on this distinction, presented in statements one through five.

> [Historical case studies] cannot lead to a general formula for the "science-technology relationship." Indeed, *such inquiries can be, and perhaps should be, conducted under complete avoidance of the terms "science" and "technology."* . . . So far we have defined "science" and "technology" in our own terms and have then tried to analyze their relationship through the course of history from our own vantage point.
> Instead, we should recognize that the concepts of science and technology themselves are subject to historical change; that different epochs and cultures had different names for them, interpreted their relationship differently, and as a result, took different practical actions.[28] (my italics)

This passage, like the quotation from Arnold Thackray cited in the introduction to this chapter, comes from the Burndy Library Conference proceedings. They bear eloquent testimony to the confusing state of the science–technology theme in TC. Before suggesting a more appropriate model for the history of technology, however, we must turn our attention to the one aspect of the science–technology discussion that can clarify the confusion and the popularity of this theme. It is the familiar debate over the formula "technology is applied science."

Is Technology Applied Science?

The applied science controversy achieved considerable prominence during the 1960s because of the furor surrounding the results of Project Hindsight published in 1966. Hindsight was a major study, eight years in preparation, undertaken by the U.S. Department of Defense to assess the importance of basic research for twenty of the nation's most important weapons systems. The study concluded that only a fraction of 1 percent of the "events" related to developing the systems could be called basic science. Ninety-one percent were technological events and nearly 9 percent were seen as "applied science."[29] The outburst of criticism following Hindsight's publication led to a second study, TRACES, which demonstrated the dependence of five recent innovations on earlier scientific research.

At the heart of the Hindsight–TRACES debate lies the question of technology's dependence on pure science. The tension surrounding it can be seen in a body of TC's discourse that approaches the level of explicit debate. A substantial number of articles adopt the hypothesis that technology is applied science. Some use this exact formula while others use language that affirms one or more of its central assumptions.

On the other hand, we find that twelve articles explicitly critique the formula and a larger group advances hypotheses contradictory to the applied science position. These articles do not quite constitute formal debate—opposing scholars are seldom named—but the focus created by the expression "applied science" allows us to identify a set of consistent assumptions supporting that position and another set contradicting it. As we shall see, these conflicting hypotheses include many of the positions already discussed in statements one through five.

Statement 6: Technology is applied science.

Mario Bunge's "Technology as Applied Science" articulates the applied science hypothesis in detail, and the consistency of his argument provides us with a pure position including all of its critical assumptions. At the heart of his argument stands a division of all human cognition into three types: scientific laws, arational rules of behavior, and "technological" knowledge.

For Bunge, pure science is a "search for new law(s) of nature" and it is conducted in the "free and lofty spirit" that liberates it from the arational distortions of custom and value preference.[30] The scientific method, which "alone can estimate the truth value of theories," can be defined as "a careful discrimination and control of the relevant variables and a critical evaluation of the hypotheses concerning the relations among such variables."[31] Bunge repeatedly asserts the power of pure science over all other forms of cognition because of its monopoly on value-free motivation and the objective validity of the controlled-variable experimental method. As such, science is the only source of truth; all other forms of cognition depend on it if they are to rise above their arational character.

Bunge divides arational cognition into three kinds of behavioral rules.

rules of conduct (social, moral, and legal rules)
rules of prescientific work (rules of thumb in the arts and crafts and in production)
rules of sign (syntactical and semantical rules).[32]

Scientific study can analyze such rules to reveal their interlocking consistencies as they occur in distinct cultural systems of belief, but in themselves they contain no objective truth whatever.[33] Bunge distinguishes all of them from what he calls "technology." In his cognitive system technology is a middle ground between science and arational rules, and it participates in the objectivity of science precisely because it is an application of science. This application operates in two ways, leading to two kinds of technological theories, which Bunge calls "substantive" and "operative."

> *Substantive technological theories* are essentially applications, to nearly real situations, of scientific theories: thus, a theory of flight is essentially an application of fluid dynamics. *Operational technological theories*, on the other hand, from the start are concerned with the operations of man and man–machine complexes in nearly real situations; thus a theory of airways management does not deal with planes but with certain operations of the personnel. Substantive technological theories are *always* preceded by scientific theories whereas operative theories are *born in applied research*.[34] (my italics)

Consider the implications of this position. Substantive technological theories cannot be articulated until science has advanced to the point where the real principles on which they are based have been understood and verified. Bunge makes no exceptions to this norm: "substantive technological theories are always preceded by scientific theories." This is consistent with Bunge's radical disjunction of all knowledge into science, which achieves truth, and nonscience, which is by definition excluded from the domain of truth. On the other hand, the operative theories of technology, since they are by definition something more rational than empirical rules of thumb, must also come from science in some fashion. Bunge explains their link with science:

> What these operative or non-substantive theories employ is not substantive scientific knowledge but the *method* of science. They may be regarded, in fact, as scientific theories concerning action, in short, as theories of actions. These theories are technological in respect of aim, which is practical rather than cognitive, but apart from this they do not differ markedly from the theories of science.[35]

Bunge's theory establishes a radical disjunction between empirical or arational rules of craftsmen before the scientific revolution and the substantive and operative theories of later technology. Because of the historically unique contribution of science, modern technological praxis has been given access to "objective reality" in a way never before possible. Thus it is that modern technology can be accurately defined as "applied science." Its theories are applications of science's laws and its rules of praxis are applications of science's method.

Although no other TC article espouses this position with Bunge's level of theoretical elaboration, we find evidence of its major presuppositions in nearly twenty articles. It is not uncommon for authors to describe technology as "applied science," although in most cases these references do not analyze what is meant by the term (appendix 3.12.1). The expression "applied science" is often accompanied by a hypothesis already considered under statement 3: that attempts to generalize about the contribution of science to technology tend to imply that science is the necessary prerequisite for modern technological progress (appendix 3.12.3). Most authors adopting this hypothesis do not argue its validity as does Bunge. It appears in their reasoning as an unquestioned assumption. Bertram Morris is typical.

> ... *technology proceeds apace with science*. If the steam engine *required* scientific knowledge far beyond what commonsense empirical knowledge could supply, today the building of satellites, radar systems, or the new weaponry demands an accumulation of *knowledge* that taxes even the best brains of engineers.[36] (my italics)

Morris is less explicit than Bunge but his language implies the same position. Modern technology is understood as a practical application of genuine knowledge, which is assumed to be scientific. The dominance of scientific knowledge over other arational and value-bound forms of human cognition is stated even more clearly in another statement by Morris.

> Man must have done things in a commonsense, empirical way before he theorized about them in a methodical, scientific way. . . . At the level of sophisticated technology, the meticulous use of diagrams and charts needs to be substituted for *feeling* or *sensibility* if understanding and control of the machine is to be effective. The computer may be a love object in science fiction; for technology it is *the product of abstract ideas of science* turned into *an automated machine that responds not to love but to coded cards*.[37] (my italics)

The value of science is seen here, as in Bunge, in its freedom from human feeling and value commitments, a freedom on which depends the objective validity of the controlled-variable experimental method. Science is historically triumphant over all earlier forms of commonsense empiricism.

Several positions that we have identified in the preceding study of science–technology language are congruent with Bunge's applied science hypothesis, although they do not necessarily imply it. The distinction between scientific curiosity and technological practicality tends to imply the applied science model when it is a universal generalization rather than a hypothesis about individual cases. Two of the three contributions of technology to science, discussed under statement 4, are consistent with the applied science position and are sometimes cited by its proponents. Bunge notes that technology poses problems for later scientific explorations, and Morris indicates that technology often creates necessary instruments for science.

Statement 7: Technology is not applied science.

The articles adopting the applied science perspective are only a small segment of TC's literature and they tend to be written by authors who are marginal contributors.[38] They are, nevertheless, of great importance for our understanding of the intellectual character of the science–technology theme in TC. Because of its absolute claims, the applied science position tends to polarize all discussions of the science–technology relationship. This polarization largely explains why so many of TC's discussions of technological knowledge occur within the science–technology frame of reference. When TC historians criticize the applied science hypothesis they are not denigrating science. Their concern is rather to establish the unique character of technological knowledge in the face of an absolute claim which, as we have just seen, radically subordinates all nonscientific cognition to the inferior status of prescientific intellectual infancy or postscientific application.

Because only a handful of authors critique the applied science position in explicit terms (appendix 3.13),[39] we cannot assert that the strong interest in the science–technology question is due to the claims of applied science alone. Still, explicit critiques indicate that "applied science" is a salient formula in TC's universe of discourse. The conflict between these two approaches, on the level of presuppositions, provides us with a very helpful framework for understanding the science–technology discourse.

We find two types of argument in TC that contradict the applied science approach. The first includes several hypotheses implying that science cannot be distinguished from other forms of cognition by a claim to objective method for achieving truth. The second offers several interpretations of technological knowledge as irreducibly distinct from science.

The most common critique of the first sort has already been seen under statement 2, which includes a set of articles arguing that the goal of science is the construction of artifacts in the form of specific theoretical models. The linchpin of Joseph Agassi's interpretation is the argument that "pure empirical data" does not exist. The cognitive limitations of scientific theories are due to the fact that scientific data must fit the theory in question before it can serve as evidence that confirms or falsifies it. We also saw that Cyril Smith's contrast between science and technology implies the same point by noting that nineteenth-century chemists ignored some available evidence because it was not congruent with the theory under construction. According to Agassi, Smith, and others like them, science *constructs* "facts"; it does not *discover* them. Therefore, like any other form of human artifact construction, the value of scientific theories must be judged by some norms other than the canon of pure and absolute objectivity.

Cyril Smith has developed a second hypothesis, unique among TC's historical articles, which implicitly critiques that applied science position from a different perspective. The hypothesis focuses attention on the interrelationships among art, technology, and science. Smith begins by noting that the encounter between an individual and the constraints of material reality was, in the earliest stages of human history, an irreducible composite of aesthetic, technological, and scientific experience.

In studying man's earliest history . . . it is difficult to separate things done for "pure" aesthetic enjoyment from those done for some real or imagined "practical" purpose. The man who selected for admiration a beautifully shaped and textured stone was yielding to a purely aesthetic motivation, but the man who molded clay into a fertility figurine was simultaneously an artist, a scientist learning to understand the properties of matter, and a technologist using these properties to achieve a definite purpose.[40]

In this early composite of art, praxis, and science the aesthetic impulse may well have been the most practical human motive.

Paradoxically man's capacity for aesthetic enjoyment may have been his most practical characteristic, for it is at the root of his discovery of the world about him, and it makes him want to live. . . . Over and over again scientifically important properties of matter and technologically important ways of making and using them have been discovered or developed in an environment which suggests the dominance of aesthetic motivation.[41]

Smith's point is that a hypothetical disjunction between aesthetic delight and either technology or science distorts the essentially human quality of both the latter activities. It will not serve to adopt a language suggesting that art is purely aesthetic, technology purely practical, and science purely rational. Such distinctions have arisen due to the recent sociological differentiation of distinct communities based on artistic, technological, and scientific professions.[42] Smith extends his analysis to modern times and, in the process, criticizes the mentality that suggests that science is a purely objective and rational activity.

It is high time that scientists admit that their experience in the laboratory is an aesthetic one, at times acutely so: the arid form of presenting their results has disguised this, and their respectable logical front often makes it invisible even to a student.[43]

Smith's argument critiques the applied science model in that he rejects any separation of the process of scientific verification (i.e., through controlled-variable experimentation) from the appreciation of beauty, an emotional experience. The argument that objective science is somehow independent of and logically prior to aesthetics—and all other nonscientific forms of conscious behavior—distorts the human *experience* of modern scientific activity as much as it does the ancient process of nonquantitative discovery of the characteristics of material reality.

Smith's aesthetic model is similar to the analysis of scientific theories as artifacts in that both attack the applied science claim to science's absolute and "objective" validity. They differ only in emphasis. To say that all scientific theories are artifacts stresses the limitations inherent in the controlled-variable experimental method itself. To say that aesthetic delight is an essential element in actual scientific practice stresses the impoverishment of any hypothesis suggesting that science is nothing more than the experimental verification process. Joseph Agassi makes a similar argument when he notes that scientific theories are algorithms but that the verification or falsification of algorithmic systems does not lead to the creation of new or competing theories. In the act of scientific creation, other forces, such as Smith's aesthetic impulse, must account for the emergence of radically new theory.[44] Taken together, these two approaches constitute a powerful critique of a central premise of the applied science position.[45]

The second attack on applied science—scholars who define technological knowledge as an irreducibly unique form of cognition—includes some articles that do not refer to science at all. They relate to the science-technology theme by implicitly contradicting the applied science position. But their significance goes far beyond a mere critique of the applied science hypothesis; it constitutes the nucleus of a new theme that includes the science–technology relationship without being constricted by its limitations.

At this point, therefore, we must shift focus. As we do so it is important to recognize that the "theme" we are about to discuss is not an explicit theme in TC at present. The four components of the new approach to technological knowledge—scientific concepts, problematic data, engineering theory, and technological skill—can all be found in TC discourse, but no one, to my knowledge, has considered them as one theme. Therefore, before exploring the new potential theme we must explain how the applied science controversy generates a question for the history of technology which calls attention to nonscientific as well as scientific dimensions of technological knowledge. One article is particularly helpful in placing the question.

In "The Structure of Thinking in Technology" philosopher Henryk Skolimowski articulates the challenge coming from the applied science model to anyone arguing for the independent status of technological knowledge.

> Many methodologists and philosophers of science insist that technology is in principle a composition of various crafts. Regardless of how sophisticated these crafts may have become, they are still crafts. It is argued that technology is *methodologically derivative* from other sciences, that it has *no independent methodoloogical status*, and that what makes it scientific is the *application* of various other sciences, natural sciences in particular.[46] (my italics)

The position described by Skolimowski directly confronts the methodological starting point of "contextual" history of technology, the methodology that has become the basis for integrating the several scholarly disciplines antedating SHOT into a single historical discipline.[47] If technology is no more than an arational craft or an application of science, then the history of technology is reduced to a chronological description of artifacts and processes which, before 1700, belong to the prescientific infancy of the human race and, after that time, are derivative of humanity's sole form of intellectual adulthood, science. The claim that science is the only objectively valid form of knowledge leaves technology "mindless," bereft of its own intellectual method. Thus specifically technological knowledge, the necessary mediator between technical design and its historical context, vanishes and with it any basis for a causal link between design and ambience.

The point is critical. If scientific progress is synonymous with the progress of knowledge then all other forms of human consciousness are destined to be overtaken by science in a triumphant process through which they are eventually governed by scientific knowledge. If, however, science is seen as a limited style whose methodological constraints and particular historical traditions make it helpful for some cognitive tasks and not for others, then science takes its place as a peer in the family of human cognitive styles. If another of these styles is technological, if technological knowledge is irreducibly distinct in its own right, then science cannot claim the role as technology's sole source of knowledge and the science–technology relationship loses its status as the only question pertaining to knowledge in technology. It becomes one of several dimensions of the question. It is for this precise reason that the science–technology relationship is an inadequate frame of reference for historians of technology.

Four Characteristics of Technological Knowledge

Before we discuss the four characteristics in detail, several points must be noted. First, it is essential that the reader keep in mind that none of the four characteristics stands alone as a complete description of technological knowledge. Each is a distinct dimension of the composite whole, but in any given article several may be understood as interrelated. It is also important to recognize that the interrelationships are not random. Scientific concepts and problematic data are frequently seen as elements of engineering theory. Engineering theory and technical skill are the two primary characteristics. Care should be taken, however, not to overstate the subordination of scientific concepts and problematic data to engineering theory. As we shall see, the contributions of these two characteristics to technological knowledge are occasionally independent of formal engineering theory. Finally, it should be noted that all four characteristics derive their unique cognitive qualities from the tension between technical design and its ambience, which defines the nature of technology itself.

Scientific concepts

Much of the analysis of science's contribution to technology found in TC has already been presented under statements 3 and 5. It is not unusual for authors to argue that some specific scientific knowledge was a necessary part of the intellectual background of an inventor or of those working on a developmental project. The value of science for explaining the underlying principles of certain long-standing technical processes is a second familiar theme. Scientific contributions to technology are also recognized by authors who discuss exchanges of knowledge between scientific and technological communities or their mutual interaction in communities that combine characteristics of both. We find, however, another perspective in TC, which interprets scientific contributions to technology in terms stressing the unique qualities of technological knowledge itself.

This perspective rests on the premise that technological knowledge, precisely as technological, is structured by the tension between the demands of functional design and the specific constraints of its ambience. Technological knowledge is unique because its design concepts are radically incomplete when they remain on the abstract level. By their very nature they must be continually restructured by the demands of available materials, which are themselves governed by further constraints of cost and time pressures and the abilities of available personnel. This tension is seen most clearly in developmental projects where, as noted in chapter 2, the governing design concept of the project is continually revised by changing goals and by specific problems encountered during the process.

When TC authors describe the role of science in developmental projects they underscore a point of major importance for understanding the science–technology relationship from the vantage point of technology. Before scientific concepts can contribute to technological knowledge they must be appropriated and restructured according to the specific demands of the design problem at hand (appendix 3.14). Thomas Smith notes that the critical inventive concepts of "Project Whirlwind" involved a restructuring of earlier scientific theory for the developmental concerns of MIT's computer project.

> An explanation of R&D progress which invokes the concept of exogenous science [i.e., science as providing "a reservoir of knowledge essential to the continuing vitality of the R&D process"] does not explain the block diagram contribution of Everett and the magnetic-core storage contribution of Forrester. These were grounded in and made possible by the logical-mathematical techniques and the understanding of phenomena which the men possessed as professionally educated engineers; originally exogenous knowledge as mathematics and electronic theory had already passed into the engineering curricula and become indigenous science in this instance.[48]

While other TC authors would argue with Smith's characterization of the theoretical engineering background of Forrester and Everett as "indigenous *science*," they would agree with his basic point that scientific concepts do not remain in their original form when they are appropriated for technical thinking.[49] This principle is seen most clearly in developmental projects, but it is also implicit in those discussions of invention that see science as a necessary part of the inventor's background. Milton Kerker's discussion of the Newcomen synthesis of previous scientific and technical traditions implies that the scientific tradition alone did not and could not result in the specific technological knowledge that Newcomen embodied in his engine design.[50]

Therefore, when the science–technology relationship is seen from the perspective of technological knowledge and when that knowledge is understood to be unique because it is shaped by the tension between design and ambience, then the role of science is exactly opposite to the intellectual superiority claimed for it in the applied science model.

> Thus, in one sense science, that is pure science, is but a servant to technology, a charwoman serving technological progress. . . . Technology is not science. By this statement I mean to say that the basic methodological factors that account for the growth of technology are quite different from the factors that account for the growth of science.[51]

Our stress on those aspects of the science–technology interaction which highlight the unique and irreducible nature of technological knowledge should not be taken as an argument that historians of technology have little regard for the contributions of disciplinary science to technology. Not all authors who adopt this perspective are concerned with the applied science hypothesis and the correlative question of technology's unique form of knowledge. Many studies of science's influence on technology assume that technical knowledge is unique and that it necessarily restructures scientific concepts for its own purposes. Their primary concern lies in the specific technological contribution of a given scientific theory of tradition. Thus Lynwood Bryant's study of the influence of thermodynamics on later engine designs implies the basic principles articulated here even though his major purpose is to assess the significant impact of thermodynamics on later technology and not to argue against the applied science model. We find, therefore, that the approach to the science–technology relationship described here can be seen from two perspectives. It stands as a refutation of the applied science model, and it provides a historiographic model for explaining the contributions of science to technology in a way that is methodologically acceptable for many historians of technology.

Problematic data

The contributions of science to technology normally take the form of theoretical concepts that operate on a high level of abstraction. By contrast, the second characteristic of technological knowledge is rooted in the specifics of technological praxis. It is not uncommon for a technical problem to arise which calls attention to an area of ignorance that had been unrecognized or considered unimportant beforehand. In such cases the practitioners either find ways to resolve the problem by gathering new data or they accept it as a limiting constraint on their activities. In either case they have contributed to technological knowledge by the very fact that they have articulated a new question. The information needed to answer such questions can be called "problematic data." This expression stresses the particular cognitive quality making such information unique, namely, the fact that it is always intended to answer a specific technical question. In TC we find three situations that call for a search for problematic data.

The most common is encountered in cases of emerging technology (appendix 3.15.1). At times the recognition of ignorance occurs in a development process. New knowledge is required before the process can be advanced through the construction of the next in a series of test models. Otto Mayr, for example, notes that Charles Porter had to conduct laborious tests to compute the magnitude of acceleration at each point of his test model's cycle before designing a subsequent generation of the engine.[52] At other times the testing process is required for a decision about introducing an already developed artifact into the marketplace. Charles H. Clark discusses the series of trials conducted by the rail industry between 1888 and 1890 which provided comparative data before a single car-coupling design was chosen from among competitors.[53] In both cases, and in similar articles, the need for new data results directly from the novelty of the technology in question.

In his study of the Page electric locomotive, Robert C. Post notes a common hazard for proponents of new technology.

> Page . . . showed an unfortunate tendency, said to be common among inventors, "to apply such tests to his ideas as are likely to exhibit their best features and genuinely to forget those tests which will exhibit their worst features."[54]

Page's need to convince congressmen to fund his project led to significant omissions in his testing procedures.

In the second situation, the lack of information is recognized not in cases of emerging technology but in problems encountered during normal use of some artifact or process (appendix 3.15.2). Sometimes the problem was a more or less catastrophic accident such as the boiler explosions on early steamboats or the fatal accident of 1877 on the Georgetown canal incline.[55] Other problems were less dramatic. Louis B. Cain, for example, notes that Ellis Chesbrough, chief engineer for Chicago's first complete sewer system, lacked detailed information about key sanitation problems such as the potential influence of floods, the amount of reservoir capacity needed, and costs of construction.[56]

Whether catastrophic or not, ignorance revealed in normal usage calls attention to a fundamental characteristic of technological knowledge. No technology is ever completely understood, even after it has been introduced into normal practice. Technological knowledge is only a partial understanding of the characteristics of real-life artifacts and processes. The same principle is implicitly operative in situations of emerging technology but it is not usually commented on in those case studies because a partial resolution of problematic ignorance is satisfactory if it leads to a successful solution of the problem at hand.

The third situation is less common in TC. A handful of articles discuss the search for new data by agencies that institutionalize the process (appendix 3.15.3). The data-gathering activities of such institutions as the Coast Survey, the Franklin Institute, the National Advisory Committee for Aeronautics (NACA), and the Bureau of Statistics combine characteristics of both previous situations. Sometimes, as in the case of NACA, testing provides information directly related to the emergence of new technology. In other cases, such as the Franklin Institute's study of boiler explosions, the institution responds to problems resulting from normal use. Every TC reference presents these institutions as responding to a need for new knowledge that has been recognized in technological practice.

All three situations share the same characteristics and together they create a perspective on technological knowledge that critiques the applied science model by revealing the uniquely pragmatic character of technological cognition. The conceptual content of the data being sought necessarily reflects the structural design of the technology that has called it forth; that is to say, data must be congruent with the specific design characteristics of its technology. This congruence is due to the fundamental character of all technology, which consists of historically specific tensions between design concepts and the limited material and societal contexts in which they exist. Since technology never exists in an abstract, ahistorical domain, it necessarily requires knowledge that is problematic in the sense defined here. Unlike scientific data which is congruent with highly abstract theoretical models, technological data is rooted in the specifics of every ambience in which it operates. Thus the commonplace activity of gathering new data can be seen as an implicit critique of the applied science model because it constitutes a form of cognition that is irreducibly distinct from science.

Engineering theory

Many TC authors discuss a form of technological knowledge that is distinct from both science and problematic data while sharing some characteristics of both. Engineering theory, whether referred to by this expression or not, has been the subject of considerable attention in TC. From a study of the more than forty references to it we can describe what appears to be a rough consensus about its meaning. An engineering theory is a body of knowledge using experimental methods to construct a formal and mathematically structured intellectual system. The system explains the behavioral characteristics of a particular class of artifact or artifact-related materials (appendix 3.16.1). Most commonly, these theories are understood to be the intellectual articulations of various branches of the engineering profession as they have developed in the United States and Europe after 1850.

From one point of view, engineering theory is more like a scientific discipline than problematic data. The governing purpose of cognitive activities in engineering theory is the creation of a theoretical artifact, namely, the theory itself. Strictly speaking, the work of formulating such a theory is only indirectly related to solving specific technical problems. Thus, in contrast to the data-gathering processes, the formation of engineering theory is more abstract and universal in its logical structure. Such theories are seen as essential for the training of professional engineers who then apply the general knowledge contained in them to the solution of specific problems.

From another perspective, however, engineering theory shares many of the characteristics that distinguish problematic data from scientific cognition. In the following passage Edwin Layton describes the difference between an engineering and a scientific theory, using solid mechanics as an example.

In solid mechanics, engineers deal with stresses in continuous media rather than a microcosm of atoms and forces. Engineering theory and experiment came to differ with those of physics *because it was concerned with man-made devices* rather than directly with nature. Thus, engineering theory often deals with *idealizations of machines, beams, heat engines, or similar devices.* And the results of engineering science are often statements about such devices rather than statements about nature. The experimental study of engineering involves the use of models, testing machines, towing tanks, wind tunnels, and the like. But such experimental studies involve scale effects. From Smeaton onward we find a constant concern with comparing the results gained with models with the performance of full-scale apparatus. *By its very nature, therefore, engineering science is less abstracted and idealized*; it is much closer to the "real" world of engineering. Thus, engineering science often differs from basic science in both *style and substance.*[57] (my italics)

We see here all of the major characteristics of engineering theory as defined above. Its intellectual focus is created by "idealizations of machines, beams, heat engines, or similar devices." Artifacts, rather than natural objects, provide its conceptual focus. The experimental procedures are intrinsically related to practical application. Thus, the content and procedures of engineering theory are, like the ad hoc style of prolematic data, intellectually structured by the demands of technological praxis rather than by the more abstract demands of a scientific discipline. TC authors have considered this form of knowledge in three contexts.

In the first context, the pressures relating to emerging technology create situations where engineering theory is particularly helpful (appendix 3.16.2). As Edward Constant has pointed out, a radically new technology has little persuasive power until it is seen to be viable. Recall that Constant distinguishes two types of "anomaly" that can result in a technological revolution. The first is called a "paradigmatic failure," in which an existing technology reveals serious limitations in its actual operation. The second, called a "presumptive anomaly," is caused by the intuition of an inventor who perceives that some new technology is possible. It is not a given, in either case, that the new technological invention will be accepted.

Anomaly resulting from either paradigmatic failure or individual intuition cannot generate community crisis; a new candidate paradigm is necessary to demonstrate the *comparative* functional failure of the conventional system. Otherwise, the alleged anomaly remains only a limiting condition to the normal technology or an eccentric speculation. . . . Few practitioners will abandon a highly successful normal technology in the absence of a *convincing* alternative.[58] (my italics)

The high cost of developing a new technology places pressure on the potential innovator to demonstrate viability before a full-scale development project will be funded. Therefore the existence of an engineering theory that can predict performance with a degree of quantifiable accuracy adds a great deal to the persuasive power of the inventive concept. John F. Hanieski discusses the predictive role of engineering theory in the context of fuselage design for supersonic flight.

A theoretical area rule existed prior to the work of Richard Whitcomb at Langley. It was, however, Witcomb's transonic area rule that *established a criterion for evaluating the relative merits of different fuselage designs*. This meant that the *expensive testing of different designs could be done more purposefully* than the prior trial-and-error approach to design.[59] (my italics)

We see here an instance of one characteristic of invention, which was discussed in chapter 2's treatment of emerging technology. The inventor must communicate the value and the nature of the new design concept to an appropriate audience before his/her idea becomes a real invention. The role of engineering theory in such cases is to provide a language for such communication.

Engineering theory often contributes to the development process as well. This is implicit in the passage from Hanieski, and it can also be seen in the analyses of Hughes, Mayr, Bryant, and others, where the frame of reference created by engineering theory is a crucial factor in the projects they discuss. These examples suggest that engineering theory is aptly described by the metaphor of language. The primary role of any language is to create a domain of communication for a group of participants. Fluency in a particular engineering domain and indeed the very existence of that domain provide a powerful resource for the costly and uncertain process by which inventive design concepts become actual innovations.

The second context in which TC authors discuss engineering theory is often related to emerging technology but it suggests another dimension. These authors stress the importance of an institutional base for the generation and communication of engineering theory (appendix 3.16.3). In most cases the base is created by an organization whose official purpose is to foster engineering theory and expertise. Typically engineering schools, professional engineering societies, or technical journals, such institutions provide a viable societal ambience in which the highly specialized communication essential for engineering theory and practice can take place. Their role in engineering theory sheds further light on the unique character of technological knowledge, which becomes clear when we consider the relationship between problematic data and engineering theory.

The definition of problematic data (specific information aimed at the solution of a particular technological problem) does not, of itself, imply any link with engineering theory. Data which solve a single problem are not necessarily applicable to any other problem. If that data falls within the domain of a general engineering theory, however, it can contribute to a much broader range of technical activity. The institutions that foster engineering theory appear to be the mediating force that has made it possible, since the mid-nineteenth century, to generalize from the individualistic knowledge learned in solving problems to theories of more universal application. We have seen that all technological knowledge is governed by the tension between the abstract universality of a design concept and the necessarily specific constraints of each ambience in which it operates. The integration of these two polarized characteristics appears to be the primary cognitive problem of technological knowledge. If problematic data were not made applicable to situations other than the problem of its origin, then it would be impossible for technological practitioners to create cumulative bodies of knowledge. Each technical problem would have to be solved as if no prior technical experience existed. Thus the role of engineering theory and the institutions that foster it can be seen as a critical factor for the creation of increasingly complex technological artifacts.

A clarification is needed here. TC authors never speak of problematic data apart from some mediating body of knowledge that allows it to be generalized. Thus the restricted definition of data used here is an abstraction. It is useful because it underscores the mediating role of engineering theory and the critically important tension between abstract and concrete knowledge. In addition, traditional knowledge—based on generalizations from experientially learned skills—serves as mediator in much the same way as engineering theory. It is omitted here so that we can focus on engineering theory. Thomas P. Hughes's case study of high-voltage electrical transmission exemplifies the role of institutionalization on several levels. The article studies attempts to understand and solve the loss of power between transmission lines at high voltage levels, a problem that came to be called "corona."

> The reverse salient appeared as a power loss between lines as transmission line voltages were raised. As Charles F. Scott, research engineer of the Engineering Department of the Westinghouse Electric and Manufacturing Company, observed: "New phenomena were liable to be presented at higher voltages and it was necessary to determine what they were, also how to meet the requirements which they impose."[60]

> The engineering experience of Scott and his colleagues led them to expect previously unrecognized problems in the new technical situation created by very high voltage. When early tests at Westinghouse laboratories revealed the corona effect, a new critical problem was identified. The Westinghouse team continued its data gathering by running tests at the Telluride gold mine in Colorado. This series of tests led to the next stage of the data search.

> Scott's ventures into an unknown world brought results of practical and theoretical interest to engineers. His experimental results, published in 1899, were admittedly limited in scope and imprecise in results,

but they brought others into a critical problem area Scott had helped greatly in defining. Within a few years after Scott published his Telluride article, a dramatic increase in technical articles on the subject of high-voltage transmission indicated that engineers had identified this salient and others were focusing upon it as a problem.[61]

By publishing his findings Scott inserted the corona problem into the larger institutional framework of the electrical engineering community. For more than two decades engineers at different research facilities continued experiments which gathered an increasing amount of data about the problem. Throughout this process the tests and their results were structured according to the intellectual canons of electrical engineering theory. Harris J. Ryan, one of the leading research specialists in the area, structured his research so that it could be of service in the actual practice of his engineering community.

The objective was not obscure: Ryan wanted to supply engineers and manufacturers tables calculated by means of his equation that would show the diameter of line conductor and the spacing that had to be used to avoid loss between conducting lines upon which specified voltages were to be transmitted.[62]

Let us consider the various ways in which this data search was influenced by the institutional contexts in which it took place. Scott's recognition of the possibility of a new critical problem depended on his previous knowledge of electrical engineering theory, and his attention was focused on such possibilities because of his position as a research engineer at Westinghouse. Thus at the outset two institutions—the electrical engineering profession and Westinghouse—contributed to the identification of the new critical problem. At the next stage of research the network created by electrical engineering journals provided an essential forum in which experimental results could be exchanged. In other words, the journals and the professional community they served allowed the knowledge gained in individual tests to become cumulative. Thus the engineering theory, which was the common language of these institutions, served to mediate between individual tests and an emerging general theory. Note, finally, that Ryan's mathematical tables, which were one of the more important bodies of data resulting from the testing process, embody the abstract-specific tension. The tables were intended to be useful for a wide range of applications but they were structured according to highly specific material characteristics such as the actual diameter of a particular metallic conductor.

Hughes's analysis is a specific example of the contrast between scientific and engineering theory as described by Layton. Hughes points out that the methodology of experimentation used in the data search is strikingly similar to the methods used in scientific disciplines. Like Layton, however, he stresses the fact that the style and the content of this data search are governed, at every stage of its development, by specific technological constraints.[63] Thus the approach to engineering theory taken by Hughes, Layton, and others is both a refutation of the applied science model and an articulation of a uniquely technological nexus between ad hoc technical problem solving and the development of generalized technical theory.

This analysis, in which engineering theory and the institutions that foster it mediate between abstract and specific elements of technical knowledge, raises the question of how this mediation took place before the emergence of nineteenth-century engineering professions. The obvious answer is that the traditions of technological skill dating to the earliest forms of technical practice served a similar role. It is not surprising, therefore, that a number of TC authors call attention to the tension between these two forms of mediation—engineering theory and technical traditions—in the nineteenth century. Articles treating this tension present a third context in which engineering theory is discussed (appendix 3.16.4). Our consideration of their approach not only concludes our discussion of engineering theory but also introduces the fourth and final characteristic of technological knowledge, technical skill.

The tension between skill and theory was recognized in the nineteenth century by proponents of two types of engineering education. In their study of the origins of the Georgia Institute of Technology, James E. Brittain and Robert C. McMath, Jr., discuss the differences between the two as seen by Robert H. Thurston, a major proponent of the theoretical approach to engineering education.

> In 1884 Thurston was stressing the distinction between "technical" schools like Stevens and "trade" schools such as Worcester. He suggested that the graduates of the technical schools would be members of a profession that would be served by trade school graduates, and he wrote that the mechanical engineer should be educated as a "designer of construction, not a constructor."[64]

Thurston's understanding of the technical–trade distinction implies a division of labor within the technological community between the theoretical expert who designs and the skilled worker who implements design. From this perspective actual technological practice could be called "applied engineering theory." This may partially explain the applied science model. Engineering theory is distinguished from skill by its quantitative methodology. The ability to codify knowledge obtained in actual practice or through testing specific materials into mathematical tables and formulas raises the precision in technical knowledge to a new level, which cannot be achieved through traditional technological knowledge. Thus the emergence of engineering theory in the nineteenth century marks the birth of a new dimension in technological knowledge. If engineering theory is understood as the application of scientific methodology and theory, and if Thurston is correct in assuming that technical skill is a simple application of engineering theory, then the applied science model is an accurate account of the changing nature of technological knowledge in recent times.

We have already argued that most TC authors reject this hypothesis on the grounds that engineering theory is irreducibly distinct from scientific theory in terms of both method and content. It remains to be seen, however, whether they understand recent engineering theory as *complementing* technical skill or as *supplanting* it. If recent engineering theory is understood as a replacement of every cognitive contribution of technical skill, then one of the major premises of the applied science model holds: recent technology would be separated from prior traditional knowledge by a radical historical disjunction. The "old" style of technological knowledge would then serve only as grist for the engineering theory mill and there would be no possible cognitive contribution for technical skill to make to recent technology in the West. In addition, we would be forced to conclude that current technological knowledge has little or nothing in common with its ancestry on the level of cognition. The earlier craftsman or inventor would be radically dissimilar to the modern engineer. This question is one of several issues that must be considered in our treatment of technical skill.

Technical skill

The tension between engineering theory and skill is only one facet of TC's treatment of technical skill. Many articles discuss skill in earlier times when its contrast with theory is irrelevant. To do justice to TC's discussion of this form of knowledge, therefore, we will consider three approaches found in the journal. As the tension between theory and skill is an underlying consideration in all three, we will consider it separately at the end.

TC authors frequently observe that technical skills are learned experientially (appendix 3.17.1). Their school is the work place. Reinhard Rürup is typical in observing that "the great English pioneers of industrial technology were not scientists but rather practical men who acquired their skills in the shops or were self-taught by trial and error."[65] Descriptions of skill fall into two classes. For some authors skill is a physiological habit, a form of learned intimacy with particular tools or machines. In an article devoted entirely to this form of knowledge, James K. Feibleman gives us the most explicit description found in the journal.

> The human individual is also a material object and if in harmony with his tools is capable of depths of understanding of them *as* material objects when he has used them long enough. Such love for particular kinds of material objects comes only through a prolonged familiarity with their use and is not confined to their form but extends more deeply into their material.[66]

Feibleman distinguishes here between abstract knowledge of form and the intimate experiential knowledge of both form and materials. His stress on the "love" and familiarity required for the second kind of knowledge differs in only one respect from Cyril Smith's argument for an aesthetic dimension in scientific and technological experience. Feibleman appears to argue for two independent forms of knowledge that can exist separately. Smith, as we noted above, suggests that the experience of even the most abstract forms of cognitive behavior involves an aesthetic or sensual dimension. In either case it is clear that technical skill requires an experiental setting for learning because it is a form of intimacy.

A more common description of experientially learned skill combines the above approach with other articles that do not discuss the question of intimacy. These articles decribe skill as the experiential basis for making technical judgments that cannot be reduced to purely theoretical knowledge. Walter R. Dornberger, for example, stresses the necessity of experience as a guide in development projects, such as the one leading to the German V-2 rocket. "A lot of technical knowledge, common sense, and experience must be expected from the chief of such an organization to guide these people, *to determine the correct moment to freeze development, and to start production*" (my italics).[67] Even in a project demanding sophisticated engineering theory, as the V-2 project clearly did, there are critical moments in which purely theoretical knowledge is inadequate. For Dornberger and others who make similar observations, it appears that theory cannot substitute for what can be called pragmatic judgment. It is inherent in the design-ambient tension of technological cognition that no technical praxis is *completely* reducible to abstract theory. An interpretation arguing that the pragmatic judgments originating in experientially learned skill are completely supplanted by engineering theory would demand such a reduction. Therefore those authors who accept the design-ambient character of technological knowledge necessarily imply that engineering theory and technical skill are two irreducibly distinct components of all technological knowledge.[68]

The pragmatic judgments referred to here are not limited to those made by managers of technological projects such as Dornberger's organizational chief. They occur in the whole array of technical processes, including that dimension of skilled work in which the worker makes judgments guiding the work process. As we shall see, TC authors do not consider every judgmental skill as irreplaceable. Even in the case of skilled workers, however, the question of how adequately they are replaced by automated machines and processes is still open.[69]

The status of skilled labor in technological praxis is, in fact, the second major approach to the question of technical skill (appendix 3.17.2). Many authors discuss situations in which skilled workers are required or are being replaced by machines or processes that begin to embody techniques previously the province of the workers. Thomas F. Glick is typical of authors who describe the need for skilled labor without reference to its replacement. Speaking of the "levelers" who were responsible for building irrigation canals in medieval Valencia, he notes that "the Valencian levelers were by no means untutored farmers but—as contemporary documents show—'persons skilled in the art of geometry and the level.' "[70]

Situations where skilled workers are replaced by more sophisticated machines or techniques are much more problematic. They are not of recent origin, as James E. Packer's study of Roman architecture indicates. His analysis of remains at the port of Ostia indicates that the city's buildings were constructed by professional labor gangs according to standardized architectural formulas. Much of the design skill was embodied in the formulas and not in the labor gangs.[71] In some cases the transfer of skill from laborer to machine was accomplished by using the skills of workers to introduce the new machines into operation. Thus Merritt Roe Smith's study of the introduction of the partially automated milling machine into early nineteenth-century arsenals indicates that the travel of skilled armorers from one arsenal to another was an essential factor in the diffusion of these machines even though they were a step in the process by which the skilled workers would eventually be supplanted.[72]

TC authors rarely discuss the labor–management tension often engendered by such transfers. Richard S. Rosenbloom's article on nineteenth-century mechanization as the locus for labor–management conflict is a rare exception. He studies an array of ideological positions relating to such central questions as class warfare, worker alienation, profit sharing, and the ethics of the factory system.[73] Another more typical article exemplifies the remarkable reticence of TC authors on this topic. In their study of the memoirs of Delaunay Deslandes—plant manager of the eighteenth-century Saint-Gobain glassworks—J. R. Harris and C. Pris refer to Deslandes's introduction of the process of casting plate glass to replace the labor-intensive practice of glass-blowing. They note that Deslandes "set himself to overthrow the old idea that the skilled work was really in blowing." He supplemented his assurances to workers that the abandonment of the blowing technique involved no diminution of their dignity with theatrical effects: "He would make a personal appearance at the moment when the casting operation began, in full dress, with a plumed hat on his head and sword at his side." Was this change of processes marked by any ill feeling on the part of workers? We hear nothing about the matter from their point of view. In the same context, however, the authors note that Deslandes was known for firm discipline: "Under him a good discipline was achieved in the factory so that officers of a nearby artillery unit when reproving soldiers for slackness "told them that they would send them to the glassworks of Saint-Gobain to be disciplined.""[74] It would seem that the significance of replacing skilled workers with "de-skilling" machines and processes cannot be fully interpreted without taking into account the perspective of the workers themselves.

Studies of this kind of tension in the work place are important for labor historians, of course, but we should stress here their importance for historians of technology. The relationship between engineering theory and technical skill is of great significance for understanding technological knowledge. If it were determined that skilled workers were a vestigial relic of premodern technology, that engineering theory were on the way to totally supplanting experientially learned skill, then instances of replacing workers with automated machines would interest historians of technology only in terms of the new automated processes and not in terms of the workers themselves. If, on the other hand, historians of technology determined that decisions made to replace skilled workers were due to nontechnical motives, such as managerial control of the work place, and if it were further determined that automation without experiential skill was a radically incomplete explanation of this type of recent technological change, then such instances would become a critically important resource for understanding the character of technological knowledge. Such a finding would demonstrate that recent engineering theory is only one component of such knowledge and that experiential skill is a cognitive dimension that is irreducibly distinct from it because it is inherent in the nature of technology itself. Clearly, the firt approach would favor the applied science model for explaining technological knowledge while the second would be another critique of that hypothesis. The lack of studies of de-skilling from the workers' perspective by historians of technology is at the same time the major reason why the question of replacing skill by automated processes remains unresolved in the field. It is also evidence that TC historians have contributed to the applied science mentality by their nearly universal failure to see worker experience as significant historical evidence. Recent studies by members of SHOT appear, however, to be beginning to fill this lacuna.[75]

The third approach to skill in TC includes many articles that discuss the cognitive processes by which experientially learned skills have been organized into bodies of knowledge that can be handed down from one generation to the next and that permit the general application of knowledge learned in individual practice. Articles discussing the organizational principles of technical skill tend to take one of two approaches. The first is already familiar from our discussion of traditional knowledge's contribution to emerging technology in chapter 2. Over the centuries empirically learned skills have been codified into rules of thumb, maxims that describe processes or general principles in nontheoretical language (appendix 3.17.3). Sidney M. Edelstein's description of the *Allerley Matkel*, a sixteenth-century handbook for spot removing and dyeing, exemplifies in one case two forms that such maxims have taken.

The *Allerley Matkel* is a small book. It was undoubtedly written for the housewife or the nonprofessional worker. Although the title says that it is concerned with the removal of spots and stains and dyeing, the recipes for cleaning are more numerous and are much more technically sophisticated than the few recipes for dyeing. This is not unexpected, for in the sixteenth century dyeing was a highly skilled art practiced and kept secret by professionals. On the other hand, the removal of stains and the renewal of garments for use was a constant problem to be solved by the common person who perhaps had only one new garment or dress each year.[76]

The recipes for cleaning epitomize a skill that was part of the public domain while the vagueness of the dye recipes reveals a more guarded tradition that was the private property of a group of professionals. Whether public or private, however, the requisite skills were articulated as descriptive recipes. The knowledge on which they were based was the result of trial-and-error experience over many generations. It is the lack of a quantitative theoretical basis for such rules of praxis that appears to be the defining characteristic for these traditional skills as defined in TC usage. In fact, the great majority of references to skill as a generalized body of knowledge stress its lack of a quantitative theoretical basis.

A handful of articles take a second approach, which may be surprising at first glance. They discuss bodies of traditional knowledge that have been codified not in verbal recipes but in mathematical formulas (appendix 3.17.4). Lon R. Shelby, for example, analyzes evidence of a mathematically structured set of rules for setting keystones in medieval arches.[77] Such mathematical rules are distinguished from later engineering theory in that the mathematics involved was elementary and they systematized a set of procedures without explaining the basis for their utility. In a study of the mathematical tables governing the design of Greek catapults, Barton C. Hacker articulates the distinction between these codifications and a theoretical model that explains behavioral principles.

> Whatever status it is accorded, however [i.e., "scientific" or not], catapult technology was clearly based on experimental investigation, its results *systematized mathematically* in the form of computed tables which allowed machines of specified characteristics to be constructed. At the same time, the mechanicians were *unable to account for their findings* or to provide them with a *theoretical framework*.[78] (my italics)

Discussions such as this suggest that the precision and replicability of mathematical formulas have been desirable for codifications of technological rules of praxis even when they were not part of formal engineering theory.

From the many TC references summarized under these three approaches we can see that technical skill is a significant theme not only for studies that predate the nineteenth-century emergence of engineering theory but also for the interpretation of recent theoretically based engineering. If the subtheme, "skill," were limited to the third approach (descriptions of nontheoretical codifications of technical praxis), it might be argued that TC authors perceive formal engineering theory as being in the process of completely replacing the older form of knowledge. It would then follow that the replacement of skilled labor by standardized machines and processes is part of the inevitable modernization of technological knowledge. But the references to experiential skill as a dimension of all technological praxis suggests a more complex interpretation. We have seen that historians of technology have, by and large, rejected the hypothesis that the cognitive dimension of technology is applied science. It seems clear from their interpretation of technical skill that they also reject the hypothesis that technological cognition is nothing but applied engineering theory. The inadequacy of both formulas derives from the same source. The tension between abstract and concrete knowledge, between design and its ambience, is the fundamental basis for the interpretation of the history of technology as it has developed in TC's universe of discourse. When this tension is seen as the defining characteristic of technology it becomes clear that the disjunction between knowing and doing, on which the applied science or applied engineering models rest, cannot serve as an adequate explanation for technological cognition. Thus, it would appear that a substantial number of TC historians interpret technological praxis as a *form* of knowledge rather than as an *application* of knowledge.

By their discussions of scientific concepts, problematic data, engineering theory, and technical skill, the authors have begun to develop a complex and provocative model. If these discussions are, in fact, the beginning of a new theme in TC, we may find that the more limited science–technology question will take its place as a subtheme within the more inclusive model. The many inadequacies of the science–technology theme appear to be directly related to the attempt to define *all* questions of technological knowledge in terms of a one-to-one relationship with science. Thus the emerging definition of technological knowledge as a unique cognitive style may explain the many significant interactions between science and technology in the very process of replacing that relationship as a major theme in TC.

4

Technology and Its Cultural Ambience

SHOT did not choose to call itself the Society for History *in* Technology, as one early wag suggested, so we can conclude that the founders were committed at least to propriety. Apart from this, however, little can be learned from the formal title except that the history of technology was the group's focus. Still, the name calls attention to the founders' decision about the title of its journal. Why didn't they choose the equally straightforward title: *History of Technology*? Where did *Technology and Culture* come from?

SHOT began at a time when the expression "history of technology" was most prominently displayed on the title page of the internalist volumes of Singer, Holmyard, Hall, and Williams. The methodology of the Singer work reflected an existing consensus about the history of technology. Disciplined attention to technical detail and to changes in design over time defined the endeavor for most practitioners. The lineage of internalist history was honorable and centuries long. For SHOT to bypass *History of Technology* as its journal title was a clear break with the past. Accounts vary slightly, but there seems to be general agreement that the word "culture" was vigorously debated due to fear that potential subscribers would understand the term as limited to the fine arts. I am told by several participants that the judgment of Lynn White, Jr., and others prevailed. *Technology and Culture* it would be, with "culture" understood in its broad anthropological sense.[1]

The title embodies the founders' desire to create a new style of history that Kranzberg and others would begin to call contextual history. As we have seen, contextual methodology has served as a paradigm for the inherent tension between technical design and its historical ambience, a tension that has served as the most helpful model for understanding TC's thematic interpretations. Chapters 2 and 3 deliberately stressed those aspects of the technological context that are immediately and obviously related to the technology side of *Technology and Culture*. In this chapter we shift our attention to the other term. How have the journal's articles interpreted the cultural ambience of technological artifacts when culture is understood in the broad anthropological sense?

A detailed survey reveals more than 850 contextual references that have some relationship to culture. They include over 250 discussions of monetary factors such as fundraising, cost constraints, and corporate technological strategies. Nearly as many references treat a broad range of governmental influences on technology such as patent procedures, funding initiatives, military policy, and regulatory constraints. Alongside these two large groups we find smaller but still substantial clusters of references dealing with the biographical background of individuals and with an array of values such as aesthetics, religion, life-style, and morality.

Although it is evident that TC authors sense the importance of culture for the history of technology, this material does not constitute shared discourse in the sense given that term in this study. There is no discernible consensus and little sign of emerging consensus about which aspects of culture should be included in contextual studies. It would be impossible to infer from present TC usage, for example, that most authors share the assumption that monetary factors must be included in technology's cultural ambience. Too many articles omit such evidence to warrant the inference. The same holds true for every aspect of culture mentioned above.

Even if such a consensus were present, we could not conclude that TC authors have begun to generate shared discourse about the design–culture relationship. Culture is more than the mere aggregate of institutional or individual behavior patterns. Strictly speaking, culture is a coherent world view, a universe of discourse giving meaning to institutions, rituals, and networks and making it possible for members of the culture to interpret reality in terms of a shared set of values and meaningful categories. Consequently, the best that could be hoped for in a detailed quantitative survey of TC references to cultural factors would be a laundry list of disparate author perceptions. If we seek an emerging consensus about the technology–culture relationship, we will not find it in this fashion.[2]

There are, however, three thematic questions in journal articles that have the potential for becoming themes of interest. Each one—technology transfer, technological determinism, and technological momentum—is broad enough to include within it the entire array of cultural factors just mentioned. They will be analyzed in detail to assess each theme's historiographical status and its suitability as an interpretative theme for the technology–culture relationship.

Technology Transfer

Technology transfer has evoked considerable interest among TC authors. Seventy-six articles refer to situations wherein some technological artifact—a material object such as a production plant or machine, or the design of a process—moves from its ambience of origin to a distinctly new one. The sole difficulty in determining which TC articles belong to this theme arises from a blurring of conceptual boundaries between "technology transfer" and the related expression "technological diffusion." Usage varies considerably. For some authors the terms are synonymous. For our purposes, however, it is helpful to distinguish the two by defining diffusion as the adoption pattern of an innovation within the culture of origin. Technology transfer, on the other hand, can be understood as multicultural, always involving the culture of origin and at least one other recipient culture.[3]

TC's reflections on transfer fall into four groups, each focusing on a different dimension of the process. The simplest is a group of articles attempting to verify claims that a given technology was transferred from one country to another. The second group includes brief discussions of the specific processes by which technologies have been carried from one culture to another. The third focuses on a more complex problem, the integration of a transferred technology into the recipient country's technological support network. Finally, a fourth group calls attention to the purely cultural tensions, as opposed to tensions related to technological support networks alone, arising from transfer. The four problems are conceptually distinct, and many TC articles deal with only one of them. But there is no logical necessity for an article to limit its consideration in this fashion, and many do not. By dividing TC's transfer discourse in this fashion we can focus our attention on this chapter's primary concern, the possibility that TC's frequent studies of transfer constitute a helpful model for interpreting the relationship between technical design and cultural ambience.

Verification of specific transfers

Of the twenty-four articles concerned with the historical accuracy of claimed transfers (appendix 3.18.1), only a handful assume this task as the central problem of the article. Thus Robert Fox's study of the origin of the fire piston in early nineteenth-century Europe is primarily intended to assess the hypothesis that the European fire piston was transferred from Asia. Conversely, Alex Roland's study of the Bushnell submarine presents evidence suggesting that the submarine was based on a design transferred from Europe to the United States and not, as had been claimed, on an original Bushnell invention.[4] More commonly, however, discussions of transfer claims occur in internalist articles whose primary purpose is tracing the historical evolution of a given technology. Thus, Claudia Kren's study of the medieval European chilinder (portable altitude sundial) includes passing references to earlier Greek, Arabic, and Indian traditions of chilinder technology, with the implication that the European version originated in transfers from one or several of these sources.[5] The similarity of transfer verifications with verification of claims to inventive originality as seen in chapter 2 is obvious. The first approach to transfer can be seen, therefore, as a part of the internalist tradition of establishing the historical ancestry of specific technologies. Such studies are not normally intended to be thematic interpretations of transfer.

Vehicles of technology transfer

Many TC discussions of transfer reveal the authors' interest in the process by which a technology moves from one place to another (appendix 3.18.2). In almost every case such discussions are passing references within the analysis of some other issue; only four articles include the question of transfer vehicles within a theoretical discussion of transfer.[6] As a group the articles identify an array of different vehicles that have served as transfer mechanisms. Thirteen articles refer to the role of skilled personnel, craftsmen, or engineers. Ten more discuss formal agreements for borrowing design blueprints, actual machines, or skilled personnel. Another eight stress the role of technical journals, institutions of higher learning, or international exhibitions for conveying technical knowledge. Transfer of technology as an element in the developmental policy of colonial powers is the subject of seven articles. Two articles treat instances of transfer in the form of technological espionage. Like the verification references above, these discussions do not in themselves involve complex thematic interpretation.

Transfer and technological support networks

When a technological artifact is carried from one culture to another, the success of the transfer depends on the ability of the recipient culture to integrate it into the infrastructure of technologies already existing there. This problem of integration has received considerable attention in TC circles (appendix 3.18.3). Forty-five articles refer to the relationship in some fashion and a handful have articulated interpretative models to explain the processes involved. Two of the most explicit, studies by Vernon W. Ruttan and Yujiro Hayami and by Nathan Rosenberg, when taken together, provide a frame of reference for understanding the major issues.

Ruttan and Hayami distinguish three phases of international transfer.

> The first phase is characterized by *the simple transfer or import of new materials such as seed, plants, animals, machines, and techniques associated with these materials. Local adaptation is not conducted in an orderly and systematic fashion. The naturalization of plants and animals tends to occur as a result of trial and error by farmers.*
>
> *In the second phase the transfer of technology is made primarily through the transfer of certain designs* (blueprints, formula, books, etc.). During this period exotic materials are imported in order to *copy* their designs rather than for their own use. . . . Machines imported in the previous phase start to be produced domestically with only slight modifications of design.
>
> In the third phase technology transfer occurs primarily through the *transfer of scientific knowledge and capacity*. The effect is to create the capacity for the production of *locally adapted* technology according to the prototype technology existing abroad. . . . An important element in the process of capacity transfer is the migration of agricultural scientists. In spite of advances in communications, diffusion of the concepts and craft of agricultural science . . . depends heavily on *extended personal contact and association*.[7] (my italics)

Several points in the model are noteworthy. In both the first and second phases, the critical transfer problem—adapting new technology to local circumstances—is not adequately carried out. In the first phase, when technological artifacts are produced in the country of origin and shipped to the recipient country, their technical design characteristics are fixed in advance and little can be done to adapt them after arrival. This inflexibility is not significantly reduced when the recipient country copies design formulas from the country of origin and produces them locally. As long as design characteristics remain fixed in the forms appropriate for the country of origin, local adaptation is ineffectual.

For a transferred technology to become an effective component in the recipient country's technological network, the transfer must occur on the third level wherein the capacity to design is itself transferred. For this reason Ruttan and Hayami stress the importance of transferring people rather than just artifacts or blueprints. The personal interactions of those skilled in designing the technology with local people who have an intimate knowledge of the technical needs and limitations of their situation is the critical factor for effective transfer.

Rosenberg's starting point is completely different, but his conclusions are similar. He begins with a discussion of technological *diffusion* within a single country, that is, diffusion from one segment of the country's industrial network to another. Using evidence from nineteenth-century American and British metalworking industries, he concludes that the key to successful diffusion lies in the capital goods sector that designs and provides equipment for a variety of industries producing consumer goods such as weapons, typewriters, and sewing machines. It is the ability of the capital goods sector to standardize production equipment for these industries that fosters rapid industrialization within the economy as a whole.[8]

Turning to the problem of industrialization in poor countries of the present day, Rosenberg draws several conclusions from his nineteenth-century examples. Although it is helpful for a country to import capital equipment, this kind of transfer deprives the recipient country of an active technical capacity and reinforces its dependence on the country of origin.

> Of course it is of enormous importance to [poor countries] to be able to import this equipment, even where the equipment is not optimally factor-biased. But if new techniques are regularly transferred from industrial countries, how will the learning process in the design and production of capital goods take place? *Reliance on borrowed technology perpetuates a posture of dependency and passivity.* It deprives a country of the development of precisely those skills which are needed if she is to design and construct capital goods that are *properly adapted to her own needs.*[9] (my italics)

Like Ruttan and Hayami, Rosenberg stresses the problem of adapting the design of imported technology to local needs. He notes that the ability to design capital equipment locally overcomes two major problems inherent in transfer situations.

> Innumerable unsuccessful foreign aid projects in the past twenty years . . . have confirmed that when modern technology is carried to points remote from its source, without adequate supportive services, it will often shrivel and die. This is partly because *the technology emerged in a particular context*, often in response to highly narrow and specific probelms. . . . But, more important, the technology functions well only when it is *maintained and nourished by an environment offering it a range of services which are essential to its continued operation.*[10] (my italics)

The attempt to insert imported technology into the recipient country's technological network often results in serious distortions where, from the viewpoint of the recipient country, the wrong tools tend to perform the wrong functions. The distortion is enhanced when the recipient country's support network in not able to service and maintain the new technology. Rosenberg concludes his interpretation with an obvious insertion. To conceive of transfer in terms of isolated bits of technology while ignoring the overall technological support network necessarily leads to failure.

> We must pay attention to the fact that the performance of individual industries will frequently depend not only on resources within that industry but on the availability and the effectiveness of *industries which stand in an important complementary relationship with it*. . . . Much of the discussion of the prospects and possibilities for technical improvement in poor countries has suffered from ignoring such interindustry relationships.[11] (my italics)

Ruttan and Hayami and Rosenberg adopt the same point of view about transfer, but they stress complementary dimensions of it. Ruttan and Hayami focus attention on the importance of human interaction, while Rosenberg underscores the role of an appropriate technological support network. In both articles the dominant problem facing transfer attempts stems from the intimate relationship between the design of every technological artifact and the technological ambience within which it functions.

The frame of reference created by these two articles is helpful when we consider the entire set of forty-five articles referring to transfer under this rubric. They can be divided into three types. The first group, including Ruttan and Hayami as well as Rosenberg, comprises nine articles that propose theoretical models. Although their approaches vary considerably in their levels of articulation, all nine discuss the tension between technical design and the recipient country's support network. The second group is made up of fourteen specific case studies, which tend to exemplify the theoretical principles enunciated by Ruttan and Hayami and Rosenberg in considerable historical detail. Dana G. Dalrymple's study of the transfer of American tractors to Russia is remarkably similar to Ruttan and Hayami's three-stage model. An early period when Russia imported American-made tractors was followed by the opening of a factory in Leningrad which made "a rather poor replica of the Fordson" copied from one of the first imported machines. Dissatisfied with the product, Russia eventually built a new facility near Stalingrad, which was designed by Albert Kahn of Detroit, constructed under the supervision of John K. Calder of Detroit, and built with American steel, using American machines and foremen. Only slowly, after the opening of the plant in 1930, did Russia develop its own design expertise and a labor force capable of running the factories effectively. Rosenberg's stress on the role of the entire industrial support network is also exemplified in Dalrymple's analysis. The actual use of the tractors was greatly hampered by the lack of an adequate network of repair facilities and by the unfamiliarity of Russian peasant farmers with machine technology.[12] The third group includes twenty articles, mainly case studies, whose references to transfer are peripheral. They also tend to follow the lines of analysis found in Ruttan and Hayami or in Rosenberg.

It seems clear that this approach to technology transfer originated in economic analysis. Six of the nine theoretical discussions and six of the fourteen specific case studies were written by economists or economic historians.[13] It is possible that the remaining eight case studies are not dependent on economic theory, but this seems unlikely when we note that none of their authors departs significantly from the transfer model presented here. For our purposes, however, the most interesting characteristic of this approach is the fact that so many of the articles make no mention of cultural tensions accompanying transfer. There is, of course, no logical necessity for transfer studies focused on support networks to ignore cultural issues. In fact, a few articles raise the question. They, along with other articles ignoring the support network, are the subject of TC's final approach to transfer.

Transfer and culture

Thirty articles include the problem of cultural tensions in their treatment of transfer (appendix 3.18.4). These articles adopt a variety of points of view and methodological styles. The first group, four articles in all, uses language suggesting that the recipient culture functions only as a hindrance to technological progress and makes no positive contribution to transfer whatever. A passage from Arthur Goldschmidt's essay on the developmental problems of "emerging countries" of the mid-twentieth century exemplifies the position.

> Technical assistance personnel find the transfer of existing technology easier in the *advanced* sectors of the dual economies of the underdeveloped world, since there is generally *no cultural barrier to be breached,* no question of resistance and receptivity, no problem of absorptive capacity. . . . Lack of capital and foreign exchange are naturally basic problems in the application of modern methods. But much of the opposition to change is *embedded in the social and economic structure of*

> *the society*. A class or group resists changes that will affect its relative position in that society. The witch doctor's objection to penicillin, the landowner's rejection of agricultural machinery, the merchant importer's opposition to indigenous industry . . . have greater relative significance [than in developed economies].[14] (my italics)

The major presupposition lying at the heart of Goldschmidt's approach is that technology is culturally neutral in the sense that technological "progress" means the same thing in every part of the world. Thus non-Western countries can only profit by the importation of new and "modern" technology. While Goldschmidt notes the need for integrating newly imported technology into the existing support network of the recipient country (his "absorptive capacity"), this integration always takes the form of the recipient country adapting to the technical constraints already operative in the support network of the country of origin. Technology transfer, in other words, is nothing more than the replication of Western technological patterns in non-Western parts of the world.

The reason Goldschmidt's "culture" operates only as a hindrance to progress is, perhaps, that he has restricted the function of culture to a remarkably oversimplified collection of vested interests. His witch doctor, landowner, and merchant importer apparently have no larger world view, no personally held values, other than their tenacious clinging to positions of status and control. What is lacking is a model of culture that takes into account the depth and passion, the subtleties and textures of every human culture. We might note in passing that Goldschmidt's assumption of a single global form of culturally neutral technological progress is exactly congruent with Mario Bunge's theory that technology is applied science. As we noted in chapter 3, Bunge's position rests on the assumption that Western technological progress since the "Scientific Revolution" is the application of a culturally neutral, value-free body of objective scientific knowledge. Thus we see in Goldschmidt's position the implications of the applied science hypothesis when applied to transfer.

The second approach to transfer and culture starts from a radically different point of view. For these authors no technology is culturally neutral and, as a result, the problem of transfer is compounded by the tension between the cultural values embedded in the design of imported technology and the culture of the recipient country. Jack Baranson articulates this perspective: "Heretofore, little attention has been paid to accommodating technology design to cultural traits; instead, emphasis has been placed upon adjusting societies to machines."[15] Robert Theobald carries Baranson's argument further when he observes that the cultural traits of "less developed countries" may well be badly needed by present-day Western society.

> It is generally accepted that economic development will require some changes in value systems. But values are not rapidly malleable; only a new generation is able to develop a radically different approach to life. As the developing countries *already possess* many of the values appropriate to a society of economic abundance, *we must re-examine the thesis that these countries must embrace the values that made abundance possible in the advanced nations.*[16] (my italics)

Goldschmidt, Baranson, and Theobald, while they represent contradictory points of view, all present theoretical reflections on mid-twentieth-century problems of transfer, and their unambiguous positions stand out from most other references to transfer and culture. Another author who discusses present-day transfer problems from a theoretical perspective calls attention to the ambiguity of the question. In his analysis of a variety of theories about the cultural implications of technology transfer, Donald W. Shriver, Jr., quotes the remarks of a young African participant in a 1966 conference of church leaders held in Geneva.

> It is all well and good for you Europeans to talk so pessimistically about the problems that technology brings to any society. But let me tell you this: we Africans want a chance to confront those problems! Do you know what technology means to us? It means that our women will no longer have to walk five miles down the road to the river for water every day; and it means that when it gets dark, instead of going to sleep, we can stay up and read a book!

Shriver follows this quotation with his own observation, highlighting the tension between the values found in contemporary Western technology and the subtle value structure of other cultures.

> Some anthropologists will reply that small changes in technology can trigger vast changes in primitive cultures. Do the Africans know what they are doing in accepting water pipes and electricity? Only in part, doubtless, and when the time comes for them to evaluate some advanced stage of technology, they may need the hypotheses of Ellul, McLuhan and company. And along with all other men facing decisions about technologies to adopt or reject, they will need careful, empirical studies of this and that technology, in this and that culture.[17]

Shriver does not offer a resolution for the tension between the cultural values of a recipient country and those embedded in technology designed elsewhere. Nor does he debate a related question, the fact that a great deal of technology transfer has been forced upon a recipient culture as part of the political or economic colonialism of Western nations.[18] Nevertheless, his perspective is significant simply because he raises the issue of culture stress with such force.

Turning to historical studies of transfer and culture, we find among all of TC's historical articles seven case studies whose authors indicate explicit awareness of the issue of cultural tensions in transfer situations. Not all articulate the relationship in the same detail. Russel I. Fries's study of British adoption of the "American system" of arms manufacture after 1854 concentrates on Britain's business and technological networks. His stress on economic and technical design factors is enhanced by brief considerations of such noneconomic factors as the British attachment to its tradition of skilled workmanship and to handworked arms for aristocratic markets. Likewise, Shannon R. Brown's economic analysis of British–Chinese interactions in the Jardine Company's silk filature at Ewo points out such noneconomic influences as the Chinese traditional style of forming business relationships to explain Jardine's problems in establishing a profitable venture.[19]

Studying the six-year British lag in adopting electrical technology, Thomas P. Hughes stresses the intricate network of British regulatory laws and conflicting political and economic interests in the 1880s. He summarizes his findings by suggesting that this array of inhibiting factors can best be understood as specific embodiments of a pervasive "zeitgeist." Without using the word "culture," Hughes comes close to cognitive anthropology's stress on a shared world view.

> The lag came out of a confluence of the legislative, technological, and economic—and something more: the *Zeitgeist*.
>
> The British "spirit of the times" as manifest in the electrical industry lag was characterized by prudence. American "go-aheadness" set off this circumspection and caution in Britain, and the Electric Lighting Act of 1882—for one thing—was a symptom of it. This legislation did not hamper the industry because it was poorly framed legislation but because it was an effort to legislate the future of an industry on a technological frontier.[20]

Harold Dorn's interpretation of Hugh Lincoln Cooper's remarkable role in transferring U.S. electrical technology to the Soviet Union concentrates on what Ruttan and Hayami called "extended personal contact and association." Cooper himself, embodying as he did the technical expertise necessary for Russia's Dnieper River hydroelectric project, was the primary vehicle for the transfer. He was engaged as chief consulting engineer not only because of his record of proven expertise in the United States, but also because his proposed design for the Dnieper project respected Russia's existing technological support systems (e.g., standard Russian rail gauge and simple construction methods suitable for a peasant work force).

Dorn goes beyond the technological support network to stress several cultural ironies at work in the transfer. Cooper not only argued for liberal free trade between communism—which he had described as "this awful disease" in earlier years—and U.S. capitalism but he also urged leniency in U.S. enforcement of trade restrictions involving Russian products of convict of forced labor. In both instances this prominent representative of U.S. capitalist industry takes a remarkably tolerant position vis-à-vis a communist and nondemocratic world view. Dorn is even more striking when describing Russia's accommodation of the Americanized, middle-class expectations of Cooper and his technical team when creating suitable living conditions for their stay in the Soviet Union.

The Soviets' high regard for Cooper's expertise is indicated . . . by the willingness of the government to build what must have been a costly colony for himself and his staff. As described by an American visitor: "Their group of brick cottages each with six rooms, kitchen and bath, central heating, hot and cold water would grace an American garden city development. Their food, imported by the shipload through Odessa, is almost exclusively American, and their sports opportunities could hardly be excelled anywhere in America . . . two concrete tennis courts, four clay courts, and a golf links.[21]

Thus, while Dorn does not articulate a comprehensive model for understanding cultural tensions in technology transfer, he calls attention to the ironies involved in the transfer-based interaction between two radically different cultures.

Barton C. Hacker's comparison of the Chinese and Japanese adaptation of Western weaponry is even more explicit about cultural tensions. Hacker suggests the following explanation for nineteenth-century China's failure to develop as a major military power in spite of its borrowing of Western weapons: "Self-strengthening [China's Western weapons policy] assumed that China could defend its traditional society against the West with Western weapons, that *the West's military technology could be detached from Western culture as a whole*" (my italics).[22]

Two recent articles present the most explicit and detailed articulations of the cultural dimensions of transfer. Bruce Sinclair's "Canadian Technology" uses civil engineering projects such as the Grand Trunk Railway, and the institutional design of Canadian civil and mining engineering societies, to exemplify his theme: "Canada imported both British and American techniques, ideas as well was engineers, then modified the imported technology to suit its own needs, *whether political, economic, or cultural*" (my italics).[23] An even more thorough analysis is found in John H. Jensen and Gerhard Rosegger's study of more than seventy years of interrelated transfers of transportation technology from Western nations to Romania. While giving full attention to the necessity of respecting Romania's existing technological support networks, the authors stress the country's unique cultural circumstances during the period.

> It would be tempting, *from the purely technological point of view*, to equate the history of transport development in the region with the seemingly similar problems confronted in many areas of the United States. But whatever analogies one might find by focusing on topography, competing commercial interests, and alternative technologies, would only serve *to obscure the essential differences in the political and cultural settings.*[24] (my italics)

The interpretation is extraordinary for TC not only because of this sensitivity to the cultural dimensions of transfer, but also for the broad array of factors included. Traditional networks of commerce, policies of nineteenth-century colonial governments, conflict between tradiitonal modes of work and Western models of labor management relations, and the intricate interactions of pride, patriotism, greed, and anticolonial anger in an emerging nation-state are all seen as directly influencing the design of bridges, rail systems, and river facilities. The article stands alone in TC as a thorough and sophisticated analysis of transferred technological design as influenced by cultural tensions.

Although seven case studies are unique for their focus on evidence of the impact of transfer on a recipient culture and the impact of that culture on the process of culture itself, they hardly constitute a body of fully developed shared discourse. Instead, they represent a historiographical situation often seen in TC—a potential area of thematic dialogue marked by cautious hypotheses limited to the case study in question. Hacker, Hughes, and Jensen and Rosegger indicate interest in developing the question as a general theme, but TC's body of literature clearly indicates that this issue is a potential rather than an actual theme in the journal.[25]

The remaining twelve articles referring to technology transfer and culture are either case studies, such as Mel Gorman's treatment of the first telegraph in India, or articles dealing with larger thematic issues, such as Nathan Rosenberg's reflections on environmental problems. All refer to cultural tensions in technology transfer but none addresses the question in great detail.

It is clear from this analysis that most of TC's seventy-six transfer articles avoid the question of culture altogether. With few exceptions the first three perspectives, comprising the great majority of TC references—verification, vehicles of transfer, and the purely technological tensions related to transfer—fail to integrate their reflections with the issue of culture. Even among the thirty "culture" articles we find three that explicitly state that transferred technology is not governed by cultural factors and another five that call for attention to culture without any research to substantiate their position. It is possible that the recent case studies by Brown, Hacker, Dorn, Sinclair, and Jensen and Rosegger may point the way for a historically based interpretation of transfer and culture, but at this point they represent only the cautious beginning of a theme.

The Debate over Technological Determinism

Modern technology . . . appears to be slipping more and more out of human control, becoming *an independent force* that holds sway over our lives and world.[26] (my italics)

In his 1974 study, "Historians and Modern Technology," Reinhard Rürup gives primacy of place, among the major technological issues of the day, to the question of technological determinism. He goes on to describe the new agenda.

For these reasons theoretical discussion about the relationship between technology and society has in the last few years ceased to revolve around questions about the meaning, nature, and "cultural value" of technology. . . . Today's discussion is concerned, instead, with the decisive question of whether and to what extent one can speak of the *"autonomy" in technology*, and, resulting from this, whether and to what extent, all other spheres of life are necessarily dependent on technology; furthermore, whether technology has developed from a means to an end in itself. (my italics)

Then, after describing the positions of Hans Freyer, Jacques Ellul, and Jürgen Habermas, Rürup makes the following observation about the role of historians in the discussion.

This debate . . . is being carried on among philosophers, sociologists, anthropologists, and economists, *with historians remaining on the sidelines.*

> This is unfortunate, for there is no doubt in my mind that the issues being raised in this context pose a challenge to historians that they can ill afford to evade. (my italics)

The passivity of historians is more than an unfortunate evasion; Rürup regrets the absence of their badly needed contribution to the debate as well.

> Philosophers and sociologists employ historical arguments in their reflections, at any rate, *though all too often in a highly abstract and much too schematic fashion*—a circumstance for which not they but rather historians are to be blamed, for *up to now they have done little to promote more concrete discussion about technological progress or about the general interdependence of technology and society.*[27] (my italics)

Rürup is not alone among TC authors in calling for a response to the question of autonomous technology. Ten of TC's twenty historiographical essays raise the issue (appendixes 2.15 and 2.16). Of these, articles by George Daniels, Eugene Ferguson, Howard Mumford Jones, Lewis Mumford, and David Joravsky are most explicit about the need for thorough historical research regarding the major premises of determinism. At first glance it appears that the status of TC's discourse confirms Rürup's critique. Thirty-eight articles address determinism directly (appendix 3.19). Of these only thirteen are historical and their references to the problem are made in passing. The bulk of the discussion is found in the journal's least representative style, "nonhistorical essays," fifteen of which refer to determinism. They are not only the largest group, when compared with contributions from the other four methodological styles, but they are far and away the most articulate. Together with the ten historiographical essays they constitute a body of literature tending to be, in Rürup's terms, "abstract and schematic." TC's three historical styles—contextual, internalist, and externalist—have not generated any explicit interpretation of determinism based on primary research.

Several other characteristics of TC's discourse should be noted before we undertake a detailed analysis of the articles. The debate is not focused on the question of whether technology exerts a determinant influence on culture. Almost all the authors accept the fact that technologies have strong impacts on cultures.[28] The debate's "abstract and schematic" character makes for talk that is more philosophical than historical. It can best be understood in terms of two major premises and three corollaries. The first premise states that autonomous technology results from a disjunction between efficiency as a norm for judging technical success and all other cultural norms. The second premise argues that technological progress follows a fixed and necessary sequence through modern history. Three corollaries follow. First, the relationship of society to technological change is always adaptation. Second, the historiographic format most congruent with deterministic and progressive technology is the "technological success story." Third, the history of technology is, in fact, an account of the gradual triumph of Western science and technology over all other forms of human praxis. Like the bulk of TC's discussion of the matter, the following analysis of these five related positions is more philosophical than historical. It is important, however, if we are to understand not only the ordinary terms of the debate, but also the reasons why TC historians do not seem to find it an inviting subject for their historical attention.

The First Premise: Determinism is based on a disjunction between efficiency and all other norms of technological success (appendix 3.19.1)

Jacques Ellul is not only the only TC author to stress the importance of this disjunction, but his expression of the premise is helpful because it is both explicit and thoroughly articulated. A set of six "characteristics" of the modern technological phenomenon is central to his approach.

1. It is artificial.

2. It is autonomous with respect to values, ideas, and the state.

3. It is self-determining in a closed circle. Like nature, it is a closed organization which permits it to be self-determinative independently of all human intervention.

4. It grows according to a process that is causal but not directed to ends.

5. It is formed by an accumulation of means that have established primacy over ends.

6. All its parts are mutually implicated to such a degree that it is impossible to separate them or to settle any technical problem in isolation.[29]

These characteristics can be reduced to two basic insights constituting the heart of Ellul's position. In the first place, the modern technological milieu is a single artificial complex of systemic components so completely dependent on one another that they must be seen as one encompassing environment. The comparison with nature is central, as is the stress on the artificial character of the technical milieu. Like nature the milieu created by technologies is a life-support system, but unlike nature it is constantly under repair because its several components have not grown organically. They have been constructed by a succession of individuals and groups of people. What is critical for Ellul, however, is not the human contribution to the several technologies but the fact that these technologies cannot exist without one another and that human life cannot exist, in the contemporary world, without the entire technological set.

The second key to Ellul's model is the autonomy of the technical milieu with respect to values and "ends" (human goals). Throughout his later elaboration of the model, Ellul stresses the fact that technological praxis has become autonomous because it is governed by an "accumulation of means" rather than in response to any cultural definition of values or goals. His point is that when a technological artifact is considered in abstraction from every norm except its function as part of a system, the only norm that is pertinent for judging its value is functional efficiency. When we assess the value of systemic components we naturally ask: "Does it work?" "Does it do what it must do so that the larger system can function efficiently?" Ellul's central premise states that this kind of functional judgment is no longer applied to systemic components as if the systems themselves were subject to other cultural norms. Efficiency has become the only norm that contemporary society uses to evaluate technological progress itself. An innovation is considered "better" than its predecessor if it performs its function more efficiently. No other cultural norms, such as morality or the dignity of the human person, are allowed to impinge on Western processes of emerging technology.[30] It should be noted that this efficiency norm is nearly identical with profitability, at least in capitalistic systems. In the West an innovation is usually judged to be efficient if it turns a profit in the marketplace.[31]

When both key insights are combined, Ellul's brand of determinism can be seen in complete form. Because individual technologies are interlocking parts of a single evolving system and because Western society has made a tacit commitment that this system should be governed only by the norms of efficiency and profitability, the ongoing evolution of the system according to its internal dynamic of increasing efficiency has become a "closed circle." It has begun not only to operate independently of nontechnical norms but to redefine all other cultural norms as functional components of the system itself. Thus a human being is no longer seen as a value in him- or herself. The value of every person lies in his/her functional contribution to the overall system.[32]

We should note that Ellul's analysis is more a vigorous critique of Western cultural values than a prediction of the inevitable triumph of autonomous technology. He follows his treatment of determinism with a detailed discussion of the "necessary conditions for a possible solution."[33] He calls for the "desacralization" of technical efficiency; efficiency must be stripped of its godlike role as sole arbiter of value and returned to the status of a more limited norm, one that is governed by the overarching commitment to human freedom. Ellul sees this desacralization as the major philosophical task facing the West today. For it to come about those in the West who perceive the threat to human freedom must communicate effectively with technological practitioners. Without such communication the practitioners will continue to construct a technical milieu that, by its powerful autonomous dynamic, tends to subvert any effort to foster human freedom.

Other TC authors hold similar positions. Scott Buchanan, W. Norris Clarke, S.J., Eugene Ferguson, and Lewis Mumford are the most explicit. All stress the trend in modern Western culture to separate functional norms from larger human goals. Buchanan states the problem as follows.

> It may be recalled that *this abstraction of technics from the arts, the rules from the actual makings*, leaves the agent–artist and his ends out. For the sake of organization, they are assimilated as merely means. Thus *we are left with an apparent automaton.*

His proposed remedy is not unlike Ellul's call for desacralization.

> If we are to deal with it competently, we must find some way of bringing back human artists and human ends into the powerful order that the technological system presents to our amazed and puzzled view.[34] (my italics)

Ferguson's starting point is different, but his principle is the same. He notes that the mentality of the engineer can tend, when it governs the writing of technological history, to treat the abstract account of design as a complete historical account.

> From the standpoint of an experienced engineer, who can appreciate many of the complexities that underlie the recounting of technical

events, the history of technology appears to be complete when it provides a chronology of events. Technical events can have a unity and fascination, often an absorbing interest, *quite apart from the society in which they occur.*[35] (my italics)

A noteworthy implication follows from this observation. Pure internalistic history of technology can operate as a cultural force, one that perpetuates the belief that technology has and continues to develop according to norms of efficiency alone. Lewis Mumford stresses the very point when he explains why the history of technology adopted an internalist style.

One of our difficulties in the past springs, I think, from the fact that the first scholars to give sufficient weight to technological changes were the anthropologists and archaeologists, who dealt with pre-literate societies for which any other data than the bare tools or weapons were largely lacking. Since all the perishable materials of technology had vanished, along with the people who had used them, the stone or pottery artifact came to be treated as *self-existent*, almost *self-explanatory* objects, influencing and characterizing the societies that used them. . . .

The fact that such durable artifacts could be arranged in an orderly progressive series often made it seem that technological change had no other source than the tendency to manipulate the materials, improve the processes, refine the shapes, make the product *efficient.* Here the absence of documents and the paucity of specimens resulted in a grotesque overemphasis on the material object, as a link in a *self-propelling, self-sustaining technological advance, which required no further illumination from the culture as a whole* even when the historic record finally became available.[36] (my italics)

These authors all understand the split between the functional-efficiency norm and other cultural norms as a uniquely modern and Western phenomenon, and all see this Western cultural option as destructive and correctable. Ferguson and Mumford, the only historians of technology among them, focus their recommendations for reform on the methodological assumptions of historians of technology themselves. Both call for a method integrating design characteristics with the nontechnological dimensions of the cultural ambience. Their call for an integration of design and ambience is echoed by Ellul, Buchanan, and Clarke, who also see it as the central reform of the deterministic cultural option. Only by reinstating technical praxis within an ambience understood to be larger than and to govern technology can the fatal determinist flaw—isolating efficiency from other norms—be corrected.

The Second Premise: Technological progress is deterministic because it advances in a fixed and necessary sequence

Robert Heilbroner's article, "Do Machines Make History?," stands alone in TC as an explicit exposition of the second premise. Early on, he states the critical question.

We begin with a very difficult question . . . whether there is a fixed sequence to technological development and therefore a necessitous path over which technologically developing societies must travel.[37]

He immediately answers it in summary fashion.

I believe there is such a sequence—that the steam-mill follows the hand-mill not by chance but because it is the next "stage" in a technical conquest of nature that follows one and only one grand avenue of advance.[38]

The true nature of technology is a "grand avenue of advance," an advance whose relationship with nature is defined as "conquest." The rhetorical tone of the sentence suggests an unqualified affirmation of the deterministic and conquering march. It would be difficult to imagine technological determinism more baldly asserted. Heilbroner realizes this and immediately qualifies the assertion by noting that "not all societies are interested in developing a technology of production" and by indicating his awareness of the "different pressures that different societies exert on the direction in which technology unfolds." These qualifications, however, do not change his central assertion of a fixed and necessary sequence except to modify it in the sense that determinism operates only "for those societies that are interested in originating and applying such a technology."

Heilbroner cites three kinds of evidence in support of his thesis: the phenomenon of simultaneous inventions, the fact that technological innovations are not "sudden leaps" but incremental advances, and the fact that technological progress "has always seemed *intrinsically* predictable." He goes on to cite as the heart of the argument two correlative arguments explaining "why technology *should* display a 'structured' history." The first is the link between advances in technology and advances in scientific knowledge, and the second is the link between advances in technology and advances in technical expertise or skill. The science–technology relationship is critical.

The first of these is that a major constraint always operates on the technological capacity of an age, the constraint of its accumulated stock of available knowledge. The *application* of this knowledge may lag behind its reach . . . but technical realization can hardly precede what men generally know. . . . Particularly *from the mid-nineteenth century to the present* do we sense the loosening constraints on technology stemming from *successively yielding barriers of scientific knowledge.*[39] (my italics)

I have emphasized certain phrases to call attention to an element of central importance in Heilbroner's theory. His argument here is exactly identical with Mario Bunge's hypothesis that technology is applied science. Every aspect of Bunge's thesis can be seen here in brief. Since the take-off period following the "scientific revolution," technology has participated in the "objective" character of science. Because it is value-free it transcends the barriers of traditional cultural values and so becomes the cutting edge of progress. Heilbroner's fixed sequence is based on science's objectivity, that is, since the scientific method is free from the hindrances of traditional, nonobjective values—Heilbroner's "barriers of scientific knowledge"—the knowledge derived from it progresses in a necessary and fixed sequence. Technological progress also follows a necessary sequence because it applies that science.[40]

Heilbroner later attempts to soften his determinism by noting that human societies influence the directions taken by technology.

These reflections on the social forces bearing on technical progress tempt us to throw aside the whole notion of technological determinism as false or misleading. Yet, to relegate technology from an undeserved position of *primum mobile* in history to that of a mediating factor, both acted upon by and acting on the body of society, is not to write off its influence but only to specify its mode of operation with greater precision.

This passage seems to suggest that technological progress does not follow a single fixed path, that technologies differ according to their cultural ambiences. This is not quite Heilbroner's intent, however, as the next sentence indicates.

Similarly, to admit we understand very little of the cultural factors that give rise to technology does not depreciate its role but focuses our attention on *that period of history when technology is clearly a major historical force, namely Western society since 1700.*[41] (my italics)

Like Bunge, Heilbroner restricts the meaning of "technology" in its fullest sense. Our ignorance of the historical and cultural origins of "technology" leads not to renewed study of links between technology and culture, but "focuses our attention" on "Western society since 1700." It is clear that scientific knowledge, understood as an objective, value-free, and progressive force in the West, is the linchpin of Heilbroner's thesis. True technology, that is to say, technology freed by Western society from the constraints of nonobjective cultural values, follows a fixed path precisely because the scientific knowledge on which it is based follows such a path. The applied science hypothesis is congruent with Ellul's premise as well. Ellul asserts that the heart of determinism lies in the Western choice to separate the internal norm of efficiency from other norms. This cultural option presupposes a commitment to the scientific method as free from every norm except its own internal canons of objectivity. Conversely, it could be argued that Heilbroner's fixed sequence could not occur without the radical disjunction of the internal norms of "progress" from all other norms. In fact, Heilbroner establishes such a disjunction as an operative condition for his argument.

> As before, we must *set aside for the moment certain "cultural" aspects of the question*. But if we *restrict ourselves to the functional relationships* directly connected with the process of production itself, I think we can indeed state that the technology of a society imposes a determinate pattern of social relations on that society.[42] (my italics)

It should be noted that Heilbroner attempts to analyze and support the basic Marxist determinism which, unlike capitalistic varieties, understands the entire science-based technological determinism as woven into a dialectical relationship with an equally deterministic evolution of socioeconomic forms (e.g., from capitalism to socialism) wherein the two determinisms mutually cause one another.[43] Heilbroner is no more successful than others adopting the orthodox Marxist position in escaping a radical determinism simply by inserting it into a dialectic of two equally rigorous determinisms. Once based on a belief that science achieves objective knowledge of the laws of nature the resulting determinism is inevitable whether in a Marxist framework or not.

Several conclusions can be drawn from our discussion of these two governing premises for the determinist position. The first is that the two are intimately related, are, in fact, correlative dimensions of a single hypothesis. The second conclusion is that this single hypothesis is none other than the applied science hypothesis already discussed at length in chapter 3. We can argue, therefore, that the critics of determinism such as Ellul, Ferguson, Buchanan, and others are also critics of the applied science hypothesis and that a proponent of determinism such as Heilbroner is also a proponent of applied science. Thus, although I have not included the applied science articles in the group referring to determinism, we can see that they are at least implicitly part of the discourse.

It is difficult to conclude with certainty that Heilbroner espouses determinism in its pure form. His frequent qualifications suggest discomfort with the pure position. Nevertheless, the key elements of the autonomous technology hypothesis appear to be deeply embedded not only in his language (e.g., "one and only one grand avenue of advance") but in his restriction of the determinism analysis to purely technical questions and to post-1700 technologies. This combination of deterministic language and occasional passing qualifications suggests that determinism may well operate in the elusive manner of an unquestioned assumption more often than on the more explicit level of fully articulated theory. This, at least, appears to be the case in the texts I will cite as examples of the deterministic corollaries that follow.

The First Corollary: The relationship of society to technological change is always adaptation (appendix 3.19.3)

If technology in the modern West is independent of cultural norms, advancing according to its own inner laws and gradually increasing its hold on society and its conquest of nature, then it follows necessarily that the response of society must be to "catch up" with technology. The most famous articulation of this position is perhaps William Fielding Ogburn's theory of "cultural lag."[44] Ogburn argues that human culture tends to lag behind the advances of science and technology so that there is always a period when societal changes are required to adapt cultural expectations and structures to the demands of a new technology. Rürup points out the importance of the thesis even as he notes Ogburn's persistent attempts to free himself from the deterministic position.

> Sociological theories were worked out in which technology was attributed a prominent role if not considered the actual driving force in the process of modern social change. The most important theoretical sociologist in this area was William F. Ogburn. Although he persistently denied that his theory of social change, formulated in 1922, was based on any fundamentally "technocratic interpretation of history," he did strongly stress *that technology and science were indeed the "great makers of social change" in the age of industrialization.*[45] (my italics)

Other TC references to societal adaptation are less elaborate, but the same mentality can be seen implicitly in passing remarks such as the following from W. E. Howland.

Nearly all immediate technical problems of overall scarcity are being solved. But new ones are arising almost as fast as the old ones disappear. And this is mainly, I believe, because *the folkways and institutions of society*—economic, political, and other—*are not changing with sufficient rapidity* to take full advantage of the improvements made possible by *advancing technology*.[46] (my italics)

Howland exemplifies two standard assumptions of determinism. On the one hand, "technology" is normally referred to in the singular as if the distinctions among differing technologies were irrelevant. This raises a serious logical question. To speak of "advancing" implies some criteria to define which way is *forward* and which is backward. But when "technology" (in the singular) advances, what criteria are available to cover the extraordinary array of different technologies? The only possible answer is, of course, that all changes are "forward." This radical optimism is Howland's second assumption. The role of society is to change quickly "to take full advantage of the improvements made possible by advancing technology." The possibility that some technological changes might not be improvements, or that some members of the society may not experience them as improvements, is not considered.

Derek J. De Solla Price uses terms suggesting the same position and hinting at a link between a deterministic technological progress and a deterministic scientific advance. After saying that science grows at a "very steady rate without even very much sensitivity to what one would suppose that societies and men desire," he goes on to ask: "Might it be that the cumulation of a technological art goes by the same sort of process and proves almost as intractable to the will of society and industry?"[47]

These texts provide us with the determinist answer to the central question of this chapter. Technology, understood as modern, Western, and science-based, is related to culture as an independent driving force demanding adaptive change from all other cultural institutions. We might note in passing that this assumption of a one-way causal relationship is an explicit part of Ellul's critique of determinism. For Ellul, the encompassing technical milieu reconstitutes all cultural values to fit within it. It is also the basis for Arthur Goldschmidt's theory of technology transfer, namely that non-Western cultures contribute nothing but hindrance to progress. Their only legitimate stance vis-à-vis technology is adaptation.

The Second Corollary: The historiographical format most congruent with technological determinism is the "technological success story" (appendix 3.19.4)

Four historians of technology and one professor of English call attention to that style of technological history chronicling technological successes. In an early essay on the challenges facing the history of technology Howard Mumford Jones critiques the "straight-line narrative of increasing success" and calls for historical research into technological failures as a corrective measure. Historians Louis C. Hunter and Robert Post provide two examples of "failure studies" in TC. Both preface their articles with comments about the importance of this perspective for the field.[48] Eugene Ferguson links his vigorous criticism of the success story motif with an observation about the elusive character of the language that often announces it in historical discourse.

> In the history of technology today I find in too many books an inadvertently stated message that sometimes overpowers the author's intended subject. In chance remarks and concluding paragraphs, the *enemies of progress are put in their places*, and contrasts are drawn between the objective success of technology and the dismally inefficient processes of society at large. (my italics)

He goes on immediately to identify such rhetoric with the determinist position.

> Such remarks and imprecations *reflect a world view* that underlies much of technological thought today . . . *the assumption that the whole history of technological development had followed an ordered or rational path*, as though *today's world was the precise goal toward which all decisions, made since the beginning of history, were consciously directed.*[49] (my italics)

Rürup echoes Ferguson in equally strong language.

> The historical category "development" was applied uncritically [in engineering histories of technology] and this resulted in an evolutionary approach to the history of technology. All too often the current fruits of technology appeared to be the *necessary outcome* of previous inventions and techniques, and one was inclined in retrospect to assume that from the very beginning the great inventions carried, as it were, the seed of all subsequent related developments, *a seed which only awaited germination according to some "natural law" of technology*. What this ap-

proach lacked above all—and for obvious reasons—was critical detachment from the object of research; in a sense it was *"company history."*[50] (my italics)

Ferguson and Rürup are particularly helpful because both articulate the conceptual link between the deterministic premise of a fixed and inevitable sequence of technological progress and the success-story format in the history of technology. Success stories are "company history" in the sense that they reinforce their own conceptual starting point. If the historian believes that technology must have developed in the way it did, it is natural that s/he will write its history as if it were inevitable. This kind of history, insofar as it becomes accepted by society at large, tends in turn to confirm the original premise of determinism and to foster its influence. Thus the success-story format can be understood as one of the elements in what Ellul understands as the tacit cultural option, made in the modern West, for a society dominated by an autonomous technological milieu.

The Third Corollary: Deterministic history of technology is an account of the gradual triumph of the West over all other forms of human praxis (appendix 3.19.5)

We have repeatedly noted that the determinist position applies only to modern Western technology. If this technology is progressing in a necessary and culture-free fashion, it is a short step indeed to the assumption that such progress is good for humankind at large and that the task of technological history is to chronicle the triumph of the West. Although very few TC authors address this issue directly, we see hints of the bias in the overwhelming popularity of Western and recent subjects among TC's research choices (table 2). Apologists for the moral value of modern technology, such as W. E. Howland and Sir Robert Watson-Watt, affirm the same presupposition as do theorists about technology transfer who argue that transfer is nothing other than "modernization" of "emerging" countries. Very few articles explicitly criticize this Western bias. William and Helga Woodruff conclude their study of the historical forces involved in Western economic and technological growth with one such criticism.

Western claims to have discovered the secrets of "self-sustained growth" have as much truth in them as the claim of the "self-made man." We talk about *the world* when often the changes we want to describe *are confined to the Western world*; we speak of the "onrush of science," whereas the lesson that emerges, even from the past century, is the conservation of progress. Advanced scientific and technological methods are concentrated in a small part of a region's or a nation's economy.[51] (my italics)

The fact that so many TC authors implicitly affirm Western technology's dominant position by their choice of research topics and by so little overt criticism of it suggests that this third corollary is an area where TC historians are most in agreement with the determinist position.

It should not be surprising that historians of technology do not find the deterministic position particularly helpful as an *interpretative* theme. By its very nature the concept of determinism is ahistorical or even antihistorical. Rürup underscores this point by quoting anthropologist Arnold Gehlen.

[Gehlen] refers to a remark Werner Heisenberg once made that is particularly interesting in the context of Gehlen's anthropological theory: "Technology [can hardly be seen as] the product of a conscious human effort to extend material power, but rather as a large-scale biological process by which the human organism's innate structures are impressed onto the human environment to an ever greater extent; *a biological process*, in other words, which *because it is just that, is beyond the reach of human control*." Gehlen himself is inclined to accept this view; he sees "genuine and apparently unlimited progress" in the tendency of our "technological-industrial culture" to conform to its own "natural" laws, and he states sharply, "Both types of process, technological progress and biological development, under the pressures exerted by the industrial system, have entered a phase of incalculable endlessness, and this alone compels us to call the era in which we live, meaning its focal point, *post-histoire*—a *post-historical phase*.[52] (my italics)

According to the autonomous technology hypothesis, technology in the modern West is posthistorical precisely because it has been cut loose from the cultural ambience, the fundamental source of historical evidence. For a determinist, the progressive development of technology resembles the unfolding of necessary implications in a mathematical formula—after the manner of a Liebnizian monad—more than it does a genuine historical process.[53] It may well be that most TC historians ignore the deterministic debate either because it is uncomfortably close to an argument for the futility of their endeavor or because they find its ahistorical character so foreign that it does not attract their interest.

It is clear that the question of technological determinism has not served as an interpretative theme for the technology–culture relationship in TC's historical articles. Far from providing a historical explanation, determinism questions the validity of the historical endeavor itself. Its popularity in Western culture over the past several centuries has tended to create a language about technological change that is directly antithetical to the contextual ideal for the history of technology, the integration of design and ambience. Indeed it could be argued that SHOT's central task has been to generate an alternative language to the "progress talk" of the autonomous technology position. Rürup's call for the participation of historians of technology in the debate may well be a call for just such a new language. How well have TC's historians succeeded in the task? Have they begun to generate a contextual interpretation of the relationship between technology and culture that lives up to their journal title? This complex and critical question will be the subject of chapter 5. Before turning to it, however, there remains one other concept, which shows considerable promise as a genuinely historical interpretation of the technology–culture relationship.

Technological Momentum as a Model for the Relationship between Technology and Culture

The fact that TC historians tend to avoid the autonomous technology model does not mean that they ignore the key issues that determinist theory tries to explain. Since the industrial revolution in the West, technology has manifested an innovative dynamism wherein the rate of technological change appears to have increased.[54] At the same time individual technologies have become increasingly interrelated in systemic networks. Recent technological changes have also generated unforeseen societal and technological consequences. The primary attraction of technological determinism may well be that it provides a model integrating these characteristics. By defining technological change as the inevitable unfolding of technical progress and by defining the resulting relationship between technology and culture as the one-way causal impact of "progress" on cultural institutions, the model provides a unified explanation of the complexities of recent technological dynamics.[55]

In recent TC articles the metaphor of technological momentum has attracted some attention as a possible thematic model that explains the same characteristics while avoiding those aspects of determinism that render it unacceptable for historians of technology. We ended our discussion of technolgy transfer by calling it "only the cautious beginning of a theme" for interpreting the technology–culture relationship. The status of technological momentum in TC is even more elusive. The expression, together with its twin, "technological inertia," appears in a handful of articles; explicit discussion of the concept is rare. Nevertheless, it appears to be an approach with considerable potential for helping historians of technology come to grips with the complexities of the technology–culture relationship. I will clarify the concept of momentum with the help of three articles that are atypically explicit in their use of "momentum" or "inertia." We will then survey TC's historical articles to identify a wide variety of interpretations that are congruent with this approach.

Momentum as an explicit model

The commonly accepted definition of momentum in mechanics is straightforward. A physical body has momentum when it moves at a given velocity and in a given direction. The amount of force needed to stop or deflect the body's movement is directly proportional to its size and speed. Thomas P. Hughes has been credited with being the first historian of technology to adopt momentum as a metaphor for interpreting the impact of technological change. David Hounshell explicitly adopts and refines the theme. Kenneth E. Bailes and Daniel J. Kevles, although less explicit than Hughes or Hounshell, use the terms "momentum" or "inertia" in ways very close to the Hughes model. Taken together, their four articles exemplify the value of the metaphor for the history of technology (appendix 3.20.1).

In his historiographic essay of 1974 Eugene Ferguson calls attention to Hughes's model while discussing the "deterministic" influence of technological systems.

> A technological system has a logic of its own that leads to a deterministic if not predictable outcome. . . . Tell the man who has invested his fortune in expensive production machinery that technology is merely a docile and neutral agent which can be picked up or dropped at will. . . .
>
> Thomas P. Hughes has given this aspect of the logic of a technological system a name: "technological momentum." He has demonstrated its importance in an article on the development in Germany of successful chemical techniques of hydrogenation. . . . Hydrogenation techniques were elevated to importance during World War I and then were applied after the war to peaceful purposes, but these could not, as Hughes put it, "absorb the creativity of engineers and chemists looking for new applications of the challenging technology they had mastered." Consequently the technologists were pushed by the technological momentum of their system into early cooperation with the Nazis in order to have support for their work on synthetic gasoline.[56]

According to Hughes's original article, published in *Past and Present* in 1969, the momentum of hydrogenation grew from a number of related factors. The technologists involved had invested pride, professional interest, and money in their technical expertise. Their original dynamism came from the pressure of defense needs during World War I, and that force continued after its originating cause had passed from the scene. It was embodied in the shared expertise and mutual vested interests of its practitioners. Because the group needed funding and continued legitimation to flourish, and because of its enthusiasm for hydrogenation, they eventually threw their weight behind the Nazis, who needed a synthetic fuel. Thus the group exerted a dual influence on Germany in the third decade of the century. On the one hand, the technical processes they had created allowed the nation to develop a fuel source independent of foreign sources, a critical asset for a country on its way to war. On the other hand, their link with the Nazis was an important legitimating factor in the party's rise to power.

Several aspects of Hughes's interpretation are significant. It is critically important to recognize that his "body in motion" was not purely technological. The technical expertise and hardware pertaining to hydrogenation were a "body" with momentum only because they had become embedded in the societal fabric of Weimar Germany. Only insofar as expertise and hardware were embodied in the personal, professional, and financial vested interests of a group of practitioners did that entire techno-cultural reality function as a moving body in the Republic. Correlatively, the impact of hydrogenation's momentum was neither purely technological nor purely cultural. It was an interlocking complex of the two. It would be overly simple to suggest that there were two discrete impacts, one technical and the other cultural. The group's political power—as a legitimating force for the Nazis—was due to its technical expertise. The synthetic gasoline, on the other hand, was an unmistakable political and cultural force because of its role as a fuel for war. For Hughes, therefore, the concept of momentum intimately binds technical and cultural realities to explain both the moving force of a given technology and its impact.

While the accumulation of ambient forces—political commitments, personal pride, financial relationships, networks of colleagues—around a technological process accounts for the force of Germany's hydrogenation momentum, it is the inherent tendency of systemic technology toward rigidity that accounts for the direction of the momentum. At its origin, any artifact has considerable design flexibility. The goals and style of those in positions to make design decisions exert great influence in the technology's early stages.[57] The further the project evolves, however, the more rigid the technology becomes. With each design decision certain potential options are closed precisely because others are taken. It was not a given, for example, that the early practitioners of German hydrogenation would come together to form a team focused on synthetic fuel. By the end of World War I, however, the specific focus of the group had become so rigid that it resisted changes in style even during the years when the external circumstances of Weimar Germany did not favor continuation of the project. This design rigidity maintained the direction of the momentum just as its practitioners sustained its force.

David Hounshell underscores the rigidity of achieved technological momentum in his contrast of Alexander Graham Bell's amateur approach to invention with Elisha Gray's professional style. He argues at the conclusion of his case study of the invention of the telephone that Bell and Gray constitute a historical paradigm in which the affective and intellectual momentum of the professional inventor tends toward personal rigidity in the face of radically new technological opportunities. "So we see a twist of what Professor Hughes has called "technological momentum," a tendency of technologists to perfect and maintain their familiar systems at whatever the cost."[58]

Daniel J. Kevles's 1975 article on the Research Board of National Security adopts the term "momentum," albeit with less thematic elaboration. Kevles uses it when discussing the struggle after World War II over control of postwar research. Writing of the Office of Scientific Research and Development (OSRD)—the spectacularly successful body of civilian scientists and engineers mobilized for national defense during the war—Kevles describes the interest of the group's personnel to reincarnate OSRD's civilian role in postwar defense research: "Without some interim agency the armed services might well lose the *momentum* and the participation of more civilian scientists in defense research" (my italics).[59] Earlier in the article Kevles had indirectly described the nature of this momentum when discussing the effect of a premature announcement that OSRD would be terminated at war's end.

Though Bush had emphasized to OSRD scientists that the plans [to terminate OSRD] would not go into effect until the defeat of Germany, his announcement of those plans had thrown his laboratories into turmoil. Staff members started looking for jobs, and contract offices wondered whether to take on new projects. Alarmed on their part, the army and navy suspected that OSRD scientists would be pulling out of the war effort too soon.[60]

For Kevles the "body in motion" that had achieved momentum was the research team itself. The pressures of war had gathered scientists and engineers together as a single unit whose force and direction were created by the combination of their own expertise and defense priorities. The announcement of plans to dissolve the unit had an immediate effect on the momentum, the cohesive dynamic force, of the group. The situation was not exactly the same as that described by Hughes. OSRD was united not by a single type of technology but by the organizational structure of the participants together with their personal commitments to national defense. Kevles makes it clear that another element in the group's momentum was a passionately held belief that civilians should help set defense priorities. When Kevles uses the term "momentum" all these factors are implicitly present as components of OSRD's dynamism.

This approach has several obvious similarities with that of Hughes. The very existence of the OSRD team was a factor to be reckoned with in the defense establishment at war's end. OSRD was one of several competing forces, several "bodies in motion," struggling for control of postwar defense research. Kevles argues that the very power of OSRD's momentum, power due to their many remarkable successes, may have contributed to their later failure precisely because that momentum led to unrealistic demands for civilian control. These demands generated a backlash and the eventual domination of research policy by the military.[61] As in Hughes's example, OSRD's momentum came from a blend of technical and cultural factors, and its impact was both technological and cultural. The specific projects selected as research priorities—during the war under OSRD and after the war under exclusive military control—were intimately linked with such factors as the desire for job security, the sense of prestige, and the commitment to civilian control by OSRD participants. The failure of their postwar efforts led to the culturally significant result of a defense research establishment largely free of civilian influence at the policy level. Once again we see that the momentum model prohibits any radical disjunction of technical and cultural elements in the "moving body" that is its focus.

In his analysis of Stalin's aviation policy in the 1930s, Kenneth E. Bailes uses the word "inertia" in much the same way that we have seen "momentum" used. The outbreak of World War II found the Soviet Union unprepared to wage air war with Germany. Technically, the Soviet problem was an air fleet oversupplied with heavy bombers and ill equipped with fighters having the speed, maneuverability, and fire power of the German Luftwaffe. Bailes's hypothesis to explain the Soviet lag adopts inertia as its centerpiece.

Western aviation experts and diplomats noted Soviet weakness of design in aviation engines and military airfames as early as 1936. . . .

This was, in fact, a period of technological revolution in world aviation, which the Soviet Union did not intiate and to which it was slow to respond.

The USSR at this time had the largest air force in the world and this fact, plus all the publicity concerning Soviet air records, *seems to have created inertia* and prevented Stalin and other responsible leaders from carrying out a basic change of aviation policy until 1939.[62] (my italics)

The successes of "Stalin's Falcons" at setting aviation speed records throughout the thirties were, according to Bailes, a deliberate government strategy to enhance national pride and distract attention from the savage purges of the period. Bailes sees them as an example of "the legitimatizing function of technology" in a rapidly changing society.[63] Although the inferiority of Soviet aircraft had been shockingly demonstrated as early as 1936 in the Spanish Civil War, the commitment to speed records continued to override a shift in defense priorities for several years. Bailes attributes this inflexibility to the style and structure of Soviet government.

The critical difference between the Soviet Union and these other countries [e.g., Britain] was the *political* style of the Stalinist leadership and the *overcentralization of decision-making powers* which existed in aviation as well as other segments of Soviet society. . . . Restricted by the norms of party discipline and intimidated by the purges, officers with unorthodox ideas were unable to change the basic direction of Soviet policy. Only Stalin and his closest associates were able to do this, and they seem to have been *captives of inertia* in the crucial period between 1937 and 1939.[64] (my italics)

In his concluding summary Bailes links the actual design characteristics of the air force (heavy bombers and expensive speed-record flights) with Stalinism itself. "A more adequate explanation [for the lag] must be sought in the *intimate connection between Stalinism* and *certain kinds of technology*, especially those projects of a proportion calculated to make the biggest publicity splash" (my italics).[65]

Bailes's use of "inertia" fits the "momentum" model precisely. The "body in motion" in this case is the Soviet air establishment. Its technical dimensions were intimately linked with Stalin's governmental style, with the intimidation of the purges, and with Stalin's need for publicity stunts for internal legitimation. The existing air technology, embodied in the larger cultural fabric, maintained its momentum from 1937 through 1939 in the face of already-known military risk. The force of this inertial body contributed to a single massive impact on Soviet society, which was necessarily both technical and cultural: the extraordinary sufferings of Soviet people under the initial German onslaught.

These four articles indicate that a thematic model based on momentum interprets the technology–culture relationship in a very different fashion than does the deterministic model. For determinism, the dynamism of technological change is exclusively technological: the unfolding, in necessary sequence, of technological progress. Its relationship with culture is a one-way causal impact. By contrast, the momentum model understands the very dynamic of technological change as the result of some technicial design embodied within a culture. Far from having an independent and necessary dynamism of its own, technical design has no force whatever unless it has become embodied in the choices and commitments of some set of cultural institutions or individuals. Thus the technology–culture relationship is intrinsically mutual.

Part of the value of the momentum model lies in the physicality of its governing metaphor, the "body in motion." The abstract necessity of technological "progress" is replaced by the historically specific dynamism of a given set of cultural choices about a given technology. The only "necessity" in the relationship arises from the limiting constraints of the technology at its present state of development. In other words, momentum interprets the dynamic of change in terms of the now-familiar tension between design and ambience. Momentum can well prove to be an apt model for explaining the powerful force of recent Western technological change and its impact on cultures, in a manner thoroughly congruent with the design-ambient tension so characteristic of TC's central methodological style.[66]

The momentum model as applicable to existing articles

While the words "momentum" and "inertia" are rarely found, TC historians frequently write of technological change in ways that fit the momentum model. As we have seen, the model stresses the "force" achieved by a technical design once it has become embodied within a cultural fabric and the "direction" of that moving force as set by the rigidities of already-completed design. The following discussions, based on journal articles, have been divided into six types of reference to the culturally embodied technical designs. Each approach identifies one way in which the dual nature of momentum can be identified in TC historical articles. None of the six approaches excludes any of the others, since design is related to its specific cultural ambience in a variety of ways. The approaches taken to it by an author depend more on the particulars of each situation and on the author's point of view than on any abstract definition of what constitutes the cultural embodiment of design.

1. The enduring nature of existing technical concepts (appendix 3.20.3). A technical concept that becomes the dominant definition of a specific type of technology, one that Edward Constant has called "a technological paradigm," can exert great influence on subsequent technology in two related ways. First, such concepts often generate further innovations within their own intellectual framework. Lynwood Bryant exemplifies this approach when writing of the role of the steam engine concept in the mid-nineteenth century.

> In this lavish variety [of ideas for new heat engines in technical journals] one species is clearly dominant, the steam engine, by which I mean the old-fashioned piston engine, and I mean the whole power plant, including boiler and condensor. *This species dominated men's minds* as well as the economy. . . . Unquestionably, the host of inventors working on heat engines throughout Europe and America found their guiding ideas as well as their models for mechanisms primarily in their experience with steam engines. They *defined their task as improving the steam engine.*[67] (my italics)

The second perspective complements Bryant's. If a technical concept dominates the imaginative and intellectual landscape within some technical tradition, it may well blind technological practitioners to the value of a radically new technology. David Hounshell stresses the exclusionary power of a dominant concept by quoting from the correspondence of Elisha Gray, a major inventor for Western Union, as he discusses the "practical" view of the telephone.

> "As to Bell's talking telegraph, it only creates interest in scientific circles, and, as a scientific toy, it is beautiful. . . ." "But if you look at it in a business light it is of no importance. . . . This is the verdict of practical telegraph men."[68]

Western Union, according to Hounshell, ignored the commercial potential of the telephone at a time when it had the opportunity to gain exclusive control over it. Even Gray, who had experimented successfully with such a device, was blinded by his "telegraphy practicality" from seeing the value of his own invention.

Both Bryant and Hounshell emphasize a single aspect of technological momentum. Dominant technical concepts tend to perpetuate themselves—to generate marginal improvements and to exclude radically new breakthrough innovations—because of the influence they exert on the imaginations and intellectual expectations of mainstream practitioners. Notice how the technology–culture relationship operates here. The dominant concept achieves cultural momentum because the pertinent agents participate in a limited cultural domain. In Bryant's case the domain is created, at least in part, by the technical journals publishing innovative articles about heat engines. For Hounshell the domain is constituted by Western Union and the community of "practical telegraph men." In the language of the momentum model, the "body in motion" is the dominant technical concept itself but only insofar as it has been accepted within some significant group.

2. The enduring nature of existing technological artifacts (appendix 3.20.4).

Once designed and constructed, many technologies do not wear out quickly; they often endure long after other possibly more desirable artifacts are available. This characteristic is a second source of technological momentum that is often noted in TC. Louis C. Hunter, discussing the remarkable longevity of relatively crude water mills, notes the significance of simple endurance for the historian of technology.

> Once the time and place of its origins have been determined and the course of its geographical dispersion traced, there is a certain tendency among historians of technology to dismiss this mill with some slight and often slighting reference to its inefficiency and limited usefulness. We are often wont in our study of the past to assign importance (and space) to events in the degree that they serve as links in a chain of development moving toward some distant and larger goal. *Yet absence of change, not getting anywhere, merely being used and useful has its own significance and interest for the historian.* The importance of the simple water-driven gristmill, and of its companion mills engaged in other tasks, is not only that it was an early forward step in man's conquest of natural forces. . . . It is also that for countless generations it has served as a tireless mechanical workhorse for common farm folk over a large part of Europe, the Middle East, and America.[69] (my italics)

The point made here by Hunter and less explicitly by many others in reference to the physical longevity of artifacts is especially significant because of the tendency in the history of technology to refer exclusively to change. The impetus and direction of technological momentum are often due to the sheer durability of a technology that exerts an ongoing cultural and technical influence simply because it is "there" and because people have come to depend on its being there. This form of momentum is partly due to simple economic forces and partly to the habits and expectations of people about their technological surround. In other words, a building, a bridge, or any other artifact may be valued in its present status because of the high cost of replacing it, because it serves a need well or because it has become part of the accepted living patterns of people in its area.

3. The enduring nature of governmental policy (appendix 3.20.5). Whether technological policy is embodied in legislation, in less formal policy statements, or simply in the mentality and vested interests of policy makers, it can be understood as a political response to existing technological realities. Kenneth Bailes's analysis of Stalin's aviation policy is a case in point. His central argument states that the specific technological choices for Soviet aircraft in the thirties were primarily political in character. John G. Burke's study of federal regulation of steam boilers in the nineteenth century is another exceptionally explicit presentation of the political dimension of technological momentum. His article traces the long struggle between proponents of boiler safety calling for strict regulation of the industry and an array of political figures convinced that the federal government had no business engaging in such regulation as well as businessmen whose financial investments in steamboats led them to oppose regulation.[70]

The point of such studies is clear. Since technological artifacts and practices are part of daily life, a government necessarily embodies some response to existing technology in its laws or in its commonly accepted understanding of governmental responses to technology in general. Whatever that political response may be, it tends to foster some kinds of technological behavior and to inhibit others. In every case the tendency of such policies to continue in effect as the political status quo makes them a cultural force contributing to some form of technological momentum.

4. The enduring nature of financial vested interests (appendix 3.20.6). The fiscal investment of a company or an individual is a powerful factor in a given technology's momentum. We have seen examples in Hounshell's study of Western Union and Burke's treatment of the lobbying efforts of steamboat owners In such cases, financial interests are understood as part of a larger complex of factors consitituting technological momentum. Hounshell, for example, attributes the blindness of telegraph men about telephones not only to financial interests but also to the power of the telegraph concept itself. In other articles, such as Seymour Chapin's analysis of corporate patent strategies or the studies of the textile industry by Irwin Feller and David J.

Jeremy, the impact of financial interests is presented as the dominant force structuring the momentum in question.[71] In either situation, once a great deal of money has been invested in a particular kind of technology the investors will change technological direction reluctantly. Such reluctance is, of course, another form of momentum.

5. The enduring nature of technological enthusiasm (appendix 3.30.7). We have seen that Hughes and Kevles stress the importance of the enthusiasm that can develop among practitioners working together on a project. Thomas Smith makes the same point in the following passage.

> The original airplane analyzer contract had been drawn up under wartime conditions late in 1944. Not so the contract for Whirlwind a year and a half later. Yet the *impetus* under which the program originally had started, the *momentum* that it had acquired, and *the keen interest in exploring the possibilities of the digital computer* all combined to perpetuate those conditions under which cost limitations were secondary in importance to engineering design considerations.[72] (my italics)

Like Hughes and Kevles, Smith notes the importance of the enthusiasm that grew among the members of MIT's Whirlwind team for the continued existence of the project. For a time after the end of the war, the Whirlwind project was a "solution (in process) looking for a problem." The urgency of a simulated cockpit for pilot training had disappeared. That Whirlwind continued to be funded, in the face of expressed doubts by the postwar Office of Naval Research, was a tribute to the enduring power of Whirlwind's momentum. No TC author argues that enthusiasm alone constitutes technological momentum, but comments like these indicate an awareness of its contribution.

6. The enduring nature of cultural values (appendix 3.20.8). It is difficult to talk about this sixth type of momentum apart from the previous five. Cultural values are implicit in each of them. Only the value-laden "embrace" of some culture is capable of giving momentum to a technical concept, an existing artifact, or a governmental policy. In like fashion, financial vested interests and technical enthusiasms are themselves expressions of influential values held by members of a culture. Thus all of the TC articles discussed under these five headings refer to the force of cultural values either in shaping new technical designs, in generating momentum for those designs, or in maintaining a momentum already established.

Nevertheless, TC historians rarely speak of these shaping or maintaining forces as cultural values. Most explicit TC references to the enduring nature of such values are passing references to culture, generally as a hindrance to some already-designed artifact.[73] Daniel P. Jones's study of the introduction of tear gas in the United States is more articulate than most. Fear of social unrest after World War I and the desire of the Chemical Warfare Service to introduce the new weapon for domestic social control confronted widespread "fear and revulsion" of gas warfare.[74] Jones portrays the tear gas struggle as a conflict between several different "bodies in motion"—the momentum within the Chemical Warfare Service to legitimate its existence (by providing one of its products for the "useful" societal needs of police departments) against the value momentum of public revulsion for chemical weapons in general. This article is typical of most TC references; the cultural opposition delayed, but did not otherwise significantly affect, the technology's introduction.

Thomas Esper's tale of the English musket's replacement of the longbow is more unusual. He demonstrates that the new technology supplanted the old in military use at a time when the longbow remained the clearly superior weapon. A cultural value shift, the decline of archery's popularity as a sport in England, was the single most important element in the transition.[75] In neither case, however, are cultural values seen as shaping the design of the technology.

The sole exception to the common TC pattern is a small group of articles dealing with the impact of local culture on new technology. Studies of "appropriate technology" normally appear in discussions of technology transfer. Jensen and Rosegger demonstrate the influence of Romanian national pride and emerging social groupings on the design of local transportation systems. Billie R. DeWalt's study of the invention of the corn seed plow in Mexico's Valley of Temascalcingo indicates that the design of the "sembradora" was the result of "social, ecological, and economic considerations" in the area.[76] In both articles, as in some other transfer studies, local culture is presented as having a momentum of its own that directly influences technological design.

Two other articles, unique in TC, study efforts during the Roosevelt administration to foresee and control long-term effects of technological change on American society. Arlene Inouye and Charles Susskind focus attention on "Technological Trends and National Policy," the 1937 government-sponsored report on technological impact. Carroll W. Pursell reviews a variety of New Deal attempts at the same kind of analysis. His concluding remarks aptly describe the difficulty of exerting cultural influence on technological systems that have already achieved momentum within a society.

> At base this [New Deal] strategy was informed by a reasonable desire to maximize the good and minimize the bad effects of that technological change which was such a signal characteristic of the 1920's and the 1930's. Unfortunately, it turned out that *not only were good and bad closely mixed in real life, but, even worse, one person's good was another person's bad.* For an administration that sought no radical reconstruction of American institutions and that eschewed any serious attempt at national planning, it was perhaps inevitable that it should avoid—and indeed intensify—the problems of technology in society.[77] (my italics)

Pursell's reflection also points to a key difficulty in explaining the impact of cultural values on the design of technologies. Cultures are complex. Even in cultures with small and relatively homogeneous populations, Pursell's observation about good and bad holds true. For the much more heterogeneous populations of industrialized cultures and, in particular, for a culture which many ethnic traditions and economic strata such as the United States, the difficulty of relating the values of those who shape the design of a technology to an overall cultural world view is significantly increased.

TC's explicit references to cultural values tend to avoid the question of how values shape design in favor of the more easily identifiable situations where some values in the culture become salient as they fight against a new technology. This is, in fact, one indication of the youthful status of the momentum theme in TC. Thomas Hughes's article in *Past and Present*, together with Eugene Ferguson's and David Hounshell's references to it, remain the only articulations of momentum as an explicit theme. More important, perhaps, it is another piece of evidence that TC historians have not yet achieved a consensus about how to talk about the relationship of technology and culture.

In this chapter we have seen thematic fragments—in the cultural studies of technology transfer and in these inchoate momentum interpretations—which are promising. In sharp contrast, we have also seen that a popular explanation for the relationship of culture and technology—the one-way causal impact of an autonomous technological force since 1700—is based on assumptions that are profoundly antithetical to a historical interpretation of technology and culture. It is a contrast we have seen before. This same repugnance, on the part of contextual historians in particular, proved to be the key that unlocked the riddle of the journal's confusing science–technology discourse. It would seem that the hypothesis of autonomous progress, either scientific or technological, plays a central role in the struggle of contextual historians to create the new historical frame of reference announced by their journal title. How to integrate "technology" and "culture," the actual designs of technological artifacts with the polyvalent and complicated dimensions of the larger human ambience? This question appears to be the critical issue facing SHOT's historians at the end of two decades.

The elusive challenge of a valid historical contextualism cannot be adequately understood without one further exploration of the autonomous progress position. Could it be that TC historians have themselves been influenced by this popular belief even as they have struggled against it? Could their difficulties in finding a language for talking about how technology and culture are related be due to a tension, within the historians themselves, between a desire to write contextual history and an inadvertent bias in favor of progress talk? It is to these subtle and delicate questions that we now turn.

5

Beyond Whig History

The danger to democracy does not spring from any specific scientific discoveries or electronic inventions. . . . The danger springs from the fact that, since Francis Bacon and Galileo defined the new methods and objectives of technics, our great physical transformations have been effected by *a system* that deliberately eliminates the whole human personality, ignores the historic process, overplays the role of abstract intelligence, and makes control over physical nature, ultimately control over man himself, the chief purpose of existence. *This system has made its way so insidiously into Western society, that my analysis of its derivation and its intentions may well seem more questionable—indeed more shocking—than the facts themselves.*

I trust that I have made it clear that the genuine advantages our scientifically based technics has brought can be preserved only *if we cut the whole system back* to a point at which it will permit human alternatives, human interventions and human destinations *for entirely different purposes from those of the system itself.*—Lewis Mumford[1] (my italics)

In January 1958 six scholars met at Case Western Reserve in Cleveland to found SHOT and begin its journal. The author of this trenchant warning about "the system" ranked high among contemporary role models for contextual technological history. Lewis Mumford's classic *Technics and Civilization* (1934) was no doubt familiar to all and very likely held a place of honor on their five-foot shelves.[2] By the late fifties the "insidious" Western commitment to a system idealizing control over rather than kinship with nature and the whole human personality had become one of Mumford's central concerns. His warning—that this commitment to a Baconian view of technics "ignores the historic process"—calls attention to one of the two intellectual challenges facing SHOT's founders and the historians who would join them in their endeavor.

By their choice of *Technology and Culture* for the journal title the founders announced their desire to create a new style of history, which Kranzberg and others would begin to call "contextual history." In 1958 aspects of the new approach could already be seen in works by scholars such as Louis Hunter, George R. Taylor, Matthew Josephson, I. B. Holley, Lynn White, Jr., Abbott Payson Usher, and Mumford himself. What was lacking was an explicit commitment by an identifiable community of scholars to a method that respected both the subtleties of technical design and the complexities of the historical ambience. It would not serve simply to juxtapose design and context. Only by integrating both dimensions would the new style begin to interpret technical designs as part of history's single woven fabric. The creation of a new historical language is challenging by definition. Contextualism was doubly difficult since it demanded technical as well as historical expertise.

The challenge of founding a discipline in 1958 was compounded by the situation that Mumford so vigorously identified. SHOT's members, like other citizens of what we commonly call Western civilization, have inherited at least some of the pervasive ideology of autonomous Western progress. Although few today hold the myth in its pure form, its influence remains powerful and debilitating, not only for the formation of scientific and technological policy but for the history of science, of technology, and of business.[3]

Mumford speaks of its origins in the Baconian definition of technics. It is helpful, however, to consider its modern birthplace in Descartes's disjunction of method from context. "Cogito ergo sum" ("I think, therefore I exist") followed from his first methodological principle: the beginning of all valid knowledge is to prove rather than to receive and trust one's existence. The unreliable experiential context must be doubted in every possible way in the analytical journey toward that certitude Descartes sought so desperately. The conceptual core of the ideology of autonomous progress adopts the Cartesian premise, a radical split of method from context. It is based on the assumption that the scientific method generates "value-free knowledge" precisely because and only insofar as it is practiced in isolation from its context. The nonscientific values of human culture and the personal feelings and values of the scientist must not critique or impinge on what Mario Bunge calls "the free and lofty spirit of pure science."[4]

Within the ideology of determinism, technology shares in science's methodological power because it is nothing more than the application of science. Just as the scientific method must not be impeded by the nonscientific bias of traditional beliefs and vested interests, so Luddite romanticism must not be allowed to hinder what Robert Heilbroner calls the "technical conquest of nature that follows one and only one grand avenue of advance."[5] Opposition to progress, scientific or technological, is invariably seen as coming from the larger cultural context in which scientific or technological methods operate to create their universally beneficial results.[6]

As Mumford observed, the myth can also be located in Bacon's theory of a "masculine" science that can conquer and master nature. The science–nature (i.e., method–context) relationship of Bacon is not one of kinship between knower and known, between method and context, but one of control and domination.[7] Finally, the myth has been linked with the laissez-faire principle of industrial capitalism. Adam Smith's invisible hand is a third method that must not be hindered by contextual critique.[8] The most succinct expression of the entire myth that I have found appears in the guidebook for the Hall of Science at the 1933 Chicago "Century of Progress" International Exposition: "Science Finds, Industry [technology] Applies, Man Conforms."[9] The point is unmistakable. Human beings must not try to shape scientific, technological, or business praxis by imposing cultural values on each one's respective method. The role of the human person or of human culture is conformity to the "higher" principles of this hierarchy.

The myth of progress poses a particularly vexing problem for historians. Belief in an autonomous progress, beginning in Europe with the "scientific revolution" and proceeding with inevitable necessity in both scientific and technological domains ever since, implies a radical disjunction of method from context and, therefore, of technological design from human culture.[10]

The point is critical for understanding the very real difficulty of SHOT's contextual goal. The split of method from context, so central to progress talk, is radically antithetical to history. Historians normally focus on some named reality—an event, a person, a political structure, a technological design—and call forth its human context to interpret what it means. It is the very essence of historical understanding to be contextual, to rescue the named reality from abstraction. In his essay on big questions in the history of technology George Daniels challenges historians of technology on this point. Commenting on the narrow definition of technology as "how things are done and made," he notes: "How things are done or made" is not, strictly speaking, a historian's question, and if we study the history of technology under this restricted definition of our subject matter, a more apt title for us would be "antiquarians of technology."[11]

Glenn Porter attributes the often-observed marginal status of the history of technology in the larger historical community to the perception of history of technology as antiquarianism. Speaking of the influence of Merritt Roe Smith's *Harper's Ferry Armory*, Porter notes first that the book is remarkable simply because it captured the attention of the Organization of American Historians.

Smith's book represents a real breakthrough for the history of technology, in the sense that it achieved *the rare feat of managing to get a lot of people who are not particularly interested in the history of technology to pay attention to a work by a historian of technology*. Perhaps even more than most historical subdisciplines, those in the history of technology write for and talk to each other. Sessions relevant to the field are relatively rare at the meetings of the major general historical associations.[12] (my italics)

Porter then explains why Smith succeeded in winning this attention.

The most important and most obvious reason why these works [Smith's *Harpers Ferry*, Wallace's *Rockdale*] have won such wide notice, then, is simply that they *successfully consider technical change in its wider social context*. This goal, long proclaimed by the more enlightened and progressive elements in the Society for the History of Technology, has actually been accomplished in these books. Their influence will be unusually strong because they embody and illustrate something desirable, rather than merely hypothesizing its possibility and its desirability. It is as if, after thirty years spent in proclaiming the feasibility and the need for safe disposal of nuclear wastes, someone actually got rid of a load of the stuff safely.[13] (my italics)

Reinhard Rürup, Gene Ferguson, and other observers echo Porter's comments, calling attention to the temptation of technology's storytellers to lapse into what Rürup calls "company history," the recounting of technological achievements as if they were the inevitable unfolding of progress.

Genuine contextualism is rooted in the proposition that technical designs cannot be meaningfully interpreted in abstraction from their human context. The human fabric is not simply an envelope around a culturally neutral artifact. The values and world views, the intelligence and stupidity, the biases and vested interests of those who design, accept, and maintain a technology are embedded in the technology itself. Contrary to Heilbroner's "one and only one grand avenue of advance," contextual history of technology affirms as a central insight that the specific designs chosen by individuals and institutions necessarily embody specific values. In summary, contextualism and the myth of autonomous progress are at odds because the progress myth assumes a value-free method divorced from every context and because it assumes one inevitable line of advance hindered only by those who violate the purity of the method and attempt to stop progress through value-laden subjective critique.[14]

It is no small matter for a group of scholars to attempt the creation of any really new historical paradigm, and indeed our common though often unreflective heritage of the Western progress ideology renders a genuine contextualism doubly difficult. It was in 1958, and remains today, a formidable task. We would do well then to conduct an overview of chapters 1 through 4 and summarize our findings about what SHOT has been up to in its first two decades by considering this tension between our received myth of progress and the challenge SHOT historians have set for themselves to integrate technology with culture. Has SHOT avoided the seductions of Whig history?[15] Has the society succeeded in creating a historical language that steps free from the antihistorical constraints of progress talk? The value of such an achievement would extend beyond SHOT's small group of scholars. By learning to talk about technical change as rooted in cultural circumstances, to speak of technologies as cultural artifacts, and by introducing this language into the larger society, they would help liberate their contemporaries and themselves from numbing passivity in the face of the inexorable march of "progress."

Several reflections on the question follow. We will begin by exploring two contradictory aspects exemplified by TC articles. On the one hand, SHOT has broken new ground and proven to be the source of a vital and badly needed contextual language about technological change. On the other hand, the society remains, in significant areas, ensnared in the web of Whig history.

The Integration of Design and Context

SHOT's early leadership fostered the new contextual approach in several ways. The following statement by Kranzberg on the importance of contextualism reveals his rationale for several related editorial practices. One was an inclusive publishing policy aimed at a catholic forum bringing divergent approaches into contact with one another. Another was the set of questions addressed to article referees.

The technologist cannot be isolated from the rest of society in his work, which is utilized by society; at the same time, the internal workings of the technology have a very definite effect upon its actual applications, usages, consequences, and the like. Therefore, no one can be completely an internalist or externalist in the history of technology, but must be something of both. *That is why the contextual approach allows for consideration of both the internalist and the externalist elements and tries to work them together.* And, of course, *that is why I asked "So what?"* [in the directions for article referees] of all the internalist pieces which I got, *in order to make the authors think of how the internal workings affected its use and appropriation by society as well as affecting later technical developments.*[16] (my italics)

Kranzberg saw the editorial commitment to the integration of internalist and externalist perspectives as one of his most important challenges. In these efforts he was part of the larger SHOT leadership who from the beginning sought to create a contextual style in their published research. The task would not be easy. Historical generalizations necessarily begin in the fragmented fashion noted by Otto Mayr: "The historian's approach is fundamentally inductive rather than deductive; it *begins* with microscopic research in hopes that the empirical data thus gathered will lead to generalizations on some higher level" (my italics).[17] Given the gradual and cumulative nature of this process, we should expect TC's early articles to reveal Mayr's "microscopic" style, and indeed we find most early contextual articles limited to cautious interpretative hypotheses of the case study in hand. As time passed, however, most contextual articles adopted a more explicit mode of shared discourse. Scholars began to refer to one another by name, either to critique or to support "generalizations on some higher level." The forum provided by TC, and no doubt by SHOT annual meetings and other less formal gatherings, has served as an arena for discourse in the three major thematic areas analyzed in chapters 2 through 4—the nature of emerging technology, the characteristics of technological knowledge, and the relationship between technology and culture. The result has been a harvest of interpretations of technological change in all three areas.

How well has this body of discourse succeeded in creating a valid contextual style? We cannot evaluate the historians' achievement adequately if we look only at the relatively few areas of explicit consensus, Mayr's "generalizations on some higher level." The youth of the discipline calls for a careful sifting of the many themes found in chapters 2 through 4. We will, therefore, briefly retrace the stages of analysis found in those chapters, calling attention as we go to each pattern that provides an opening in the direction of a genuine contextual integration of technical design and cultural ambience.

Studies of emerging technology (chapter 2)

Given the status of the emerging technology theme as the area of greatest creativity in TC discourse, it is not surprising that several significant subordinate themes foster the contextual perspective.

Invention. Two dimensions of TC's treatment of the process of invention open the possibility of understanding the link between design and cultural context. The first is the observation, made by several authors, that a design concept cannot be considered a true invention unless its significance has been recognized by the inventor and, more important, has been successfully communicated to an appropriate audience (appendix 3.1.2). In the eleven pertinent articles the need for congruence between inventive design and its audience is generally treated in a limited frame of reference where inventor and audience share the same basic world view and where cultural congruence is presumed.[18] Nevertheless, the approach is open to the possibility of interpreting the success or failure of technical designs in a more explicit cultural context. If a potential inventor must demonstrate the value of a new idea to an audience for recognition and validation, it follows that cultural congruence—or dissonance—is part of the story of the new invention.

In like fashion, the thirteen articles that discuss personality and motivation of the inventor are open to the idea that inventive design is a reflection of the inventor's cultural perspective (appendix 3.1.4). Though not often noted in TC, a cultural world view is part of every inventor's personal style. Both perspectives break with the ideology of autonomous progress by the very fact that inventive designs are shown to be dependent on a context larger than their purely technical dimensions. In that sense, both make a small but real contribution to the emerging contextual approach.

Development. Two common aspects of the developmental process of emerging technology foster a contextual view. The need for project leaders to lobby for funding, noted in nine articles, demonstrates the same requirement for congruence between project design and funding constituency just observed in invention studies (appendix 3.2.2). More to the point, TC's frequent analysis of a project's goal definition and its role in design changes call attention to the malleability of design concepts (appendix 3.2.5). Far from being the inevitable unfolding of an autonomous dynamic of progress, developmental models are shown in this perspective to be remarkably flexible. The point is not that design constraints do not affect development. It is, however, the inextricable relationship between such constraints and the often shifting priorities of project leaders and their funding constituencies that give TC's developmental model its historical sophistication.

Innovation. Innovation studies regularly stress the influence of economic constraints on design decisions (appendix 3.3.2). In most cases these discussions are situated in the context of Western capitalism and, as noted below, there is remarkably little critical reflection on capitalism itself in the journal. Nevertheless, it would be a relatively short step from the present consensus about the influence of capitalist motives on successful innovations to the further question of how the capitalist world view influences technical design.

The technical ambience: technological support networks, technical traditions, and systems models. Two of TC's three models for the technological ambience have little to say about integrating design and cultural ambience. The relationship between new technology and the existing technological support network is, by definition, limited to design congruencies (appendix 3.4.1). The systems model can even be seen as a subtle argument in favor of the autonomous progress model if carried to extremes (appendix 3.6.1–8). Nevertheless, neither approach necessarily implies the ideology of progress, and both can serve as part of a contextual style. It would be just as much a violation of the design-ambient integration to ignore design congruences as it would be to ignore design-culture congruences. In any body of scholarship that attempts to do justice to technological praxis we should find, as we do, serious and detailed consideration of the subtleties of design interaction.

TC's treatment of technical traditions is more explicitly related to a contextual approach. Traditions are understood to be embodied in communities whose members are admitted on the basis of the judgment of their peers. Such judgments are influenced by a larger-than-technical set of norms. Thus, although we find little discussion of cultural dissonance of congruence in TC treatments of technical traditions, the very concept of traditions as communal is open to the broad cultural approach of design-ambient integration (appendix 3.5.1–10).

Science vs. technology relationships and the nature of technological knowledge (chapter 3)

Discussions of science–technology relationships frequently critique the hypothesis that technology is merely the application of scientific knowledge. In the process TC authors have begun to talk about technological knowledge as a unique form of cognition. Its uniqueness derives from its necessary integration of design thinking with concrete judgments about complex historical situations. Here, perhaps more than elsewhere in the journal, we find an unequivocal affirmation of technological praxis as rooted in an integration of design and ambience. The frequent observation that engineering theories are distinctive because they are structurally oriented toward concrete praxis is the basis for the related conclusion that technological practitioners must rework scientific concepts in order to use them. More to the point, the distinction between engineering theory and experientially learned skill situates technological knowledge within the traditions of praxis that are part of the larger cultural ambience.

The critical point, in this evaluation of TC's creation of a nonprogressive model of technological change, is that TC authors do not portray technological thinking and praxis as operating in an abstract vacuum. This approach leads to the inescapable conclusion that technology by its very nature is rooted in the human fabric. As in the case of the emerging technology models cited above, it can be argued that TC authors have created a model of technological cognition that is open to the integration of design and ambience in every sense, even though relatively little has been done to develop the cultural implications inherent in their approach.

Design and cultural ambience (chapter 4)

The first two themes focus more on technology, or technological knowledge, than on "culture." When we consider how TC authors have handled the other side of the journal's title, we find a more complex situation. Chapter 4 discussed the still inchoate state of TC's voluminous references to "culture" and "society." It should be recalled, however, that a handful of articles have begun to raise the question of cultural influences in design processes in the two areas of technological transfer and technological momentum.

Technology transfer. Seven articles, most of them very recent, call attention to cultural tensions in technology transfer (appendix 3.18.4.C). The inherent rigidity of technical designs once they have become successful in one culture, together with the dissonance that can be observed when they are transferred to a radically different culture, are illustrations of every technology's cultural relativity. These articles are not yet a body of explicit discourse, and much remains to be done before transfer and culture becomes a mature theme. Still, the task has not been ignored altogether.

Technological momentum. The importance of this relatively recent and underdeveloped theme as a critique of autonomous progress lies in the momentum metaphor itself (3.20.1). To speak of a technological design as having achieved momentum implies that momentum must be something more than the design itself. "Achieving" momentum implies societal acceptance of and later adaptation to the constraints of the design. The contrasting case, even if not mentioned, is that of designs that do *not* achieve a place within the societal fabric. By naming this process of acceptance and adaptation, and by calling attention to the societal force that it gives to the design, authors adopting the metaphor evoke an understanding of technical design as intimately part of culture. It is impossible to use the metaphor without implying the related questions of why the design achieved momentum and how that momentum affects later societal developments. It is true, of course, that the figure can be used without calling attention to these cultural issues, but the model clearly opens the door to a cultural interpretation of successful technologies. Given the dangers of focusing too much attention on technological success stories, the momentum model would seem to offer great promise as a way of talking about such successes without falling into the autonomous technology trap.

The relationships among three historiographical styles: internalist, externalist, and contextual

To conclude this summary let us reflect on some earlier observations about the entire body of TC's historical articles. Contextual history takes as its central task the integration of design characteristics with both the cultural and technical ambience. The most popular and creative thematic discourse among contextual historians in the first two decades deals with emerging technology. Finally, we see that most contextual articles are detailed case studies rather than general surveys. These three characteristics imply a contextual critique of autonomous determinism on several counts.

Determinism asserts, in the first place, that technological change occurs within a purely internal frame of reference, that norms of internal and functional efficiency are the only ones governing it. But by situating their studies of emerging technology within the larger cultural ambience and by searching out the specific historical factors—both technological and nontechnological—that influence a given technical change, contextual historians implicitly argue that technological change does not occur in an internalist vacuum. More important, determinism asserts that modern technological change follows a fixed and necessary sequence. But the very act of identifying historically specific ambient factors and assessing their influence on the design characteristics of a new technology necessarily implies that the resulting design need not have taken the form that it did. There would be no point whatever in this contextual labor if the technical outcome were the same in any case. Thus the set of TC articles attempting to identify ambient influences during the processes of emerging technology constitutes a powerful, historically based critique of determinism.

The same critique is implicit in the entire body of externalist articles. As we observed in chapter 1, the comparatively recent emergence of externalist methodology is a result of the cumulative influence of contextual studies in TC. Contextual research, by its continual studies of the technological ambience, tends to create a new historical concept—"the technological ambience" itself—as a subject for historical research. This new concept appears to be the basis for TC's increasing interest in externalist research. Thus these articles tend, by their very existence as a distinct methodological style, to critique determinism. Detailed research into the external ambience of technology would be irrelevant if technology were radically independent of that ambience. If externalist articles tended exclusively to study the impact of technology on society it might be argued that they are nothing more than a part of determinism's technological success story format. But in fact the great majority of externalist articles studies dimensions of the technical ambience that influence technological change rather than the other way around (appendix 2).

It should not be assumed from this line of argument that TC's internalist articles necessarily imply a deterministic position. Although they embody the disjunctin of design from ambience, they play a complementary role in the journal's overall body of research. Effective contextual history calls for historical precision about the design characteristics of each type of technology. If all of TC's historical articles fell into the internalist or the externalist group, it could be argued that the journal had achieved no integration at all, that the two approaches were proceeding apace with little interaction. We find, however, that both approaches have influenced, and been influenced by, the much larger group of contextual articles (see chapter 1). Thus the relationships among the three styles are further evidence that SHOT has begun to achieve the contextual objective.

Finally, it is noteworthy that a significant number of TC's contextual articles show classic signs of internalist discourse—tightly focused scholarly exchange and a central interest in artifact design. They suggest that TC's overall commitment to contextualism has invited some internalist historians to expand their methodological horizons to include ambient factors. This is further evidence that the journal's dominant methodological commitment tends away from a radical disjunction of design and ambience.[19]

These gleanings from TC articles provide substantial evidence that SHOT has begun to live up to its journal title, that TC authors are already sharing the beginnings of a new language about technological change that is at once technically careful and liberated from the excesses so often found in progress talk. It is an achievement to be proud of, but it is not the only answer that can be given to the contextual question. Seen from another perspective, TC's articles provide evidence that SHOT has not succeeded in freeing itself from Whig history.

Signs of Whiggery

From 1960 to the present internalist articles hold a consistent share, roughly 17 percent, of TC's articles. It might be argued that this is clear evidence of SHOT Whiggery, since internalist research fosters an image of the history of technology as disinterested in and disconnected from nontechnical context. On the other hand, it can be argued that the commitment of internalists to remain active in SHOT indicates their openness to the broader questions of the society. More important, however, the internalist labor—detailed study of technological designs and their changes over time—is as essential as the detailed study of technological contexts for a valid contextualism.[20] On both counts, therefore, care should be taken before attributing too much significance to the internalist presence as an argument for Whiggery in TC's universe of discourse. Let us turn to a more serious matter, the striking contrast between thematic richness in some areas and relative linguistic poverty in others.

Imbalances in thematic discourse

While TC's interpretations of emerging technology and of technological knowledge are rich with nuances that critique the mythology of progress, the absence of an equally rich interpretation of the cultural side of our technology and culture mandate is troubling. If a suitable integration of design and context is a central task for the field, what are we to make of these strata of thematic development?

The thematic imbalances underscore a point of considerable importance. The only consensus in TC about what constitutes the technological ambience is so broad as to be platitudinous. Contextual historians might agree on the following definition: The technological ambience includes any historical particulars not immediately part of the internal design of the technology in question. Such a broad definition includes a whole host of potential ambient factors. Thus other technologies in the support network, preexisting communities of technical praxis, the several components of technological knowledge (scientific theories, engineering theory, problematic data, and experiential skill), and the entire array of cultural factors such as economic constraints, corporate institutions, governmental and political matrixes, personal and biographical details, and cultural values are all members of the set. The fact that so vague a definition of ambience is all that is possible in the TC forum suggests that much remains to be done before a mature consensus can be reached about the relationship between design and ambience.

We have noted, in chapters 2 and 4 particularly, that the sophisticated integration of design and ambience found in themes relating to emerging technology primarily deals with existing technological support networks and technical traditions. The popular systems model, on the other hand, tends to subsume ambient factors as functional components of large-scale systems rather than treating them as exogenous contextual factors.[21] Thus it could be argued that TC's well-wrought discourse about emerging technology, when contrasted with the poverty of similar discourse about cultural factors, is an indication of the journal's Whiggish tendency to concentrate on technological factors and to treat the larger ambience as a peripheral concern.

Chapter 3 called attention to the popularity of the scientific ambience for many TC historians. While we have argued that several dimensions of TC's treatment of this issue serve as powerful critiques of the Whig position, argument can be made for the opposite position. The contrast between TC's serious pursuit of the science–technology relationship and the lack of a similar level of concern for the cultural ambience is a subtle encouragement for the Whig position. The myth of autonomous technological progress is, after all, part of a larger ideology—the triumph of scientific and technological progress in the West. From this point of view, the absence of a balancing body of discourse about other cultural relationships to technical design suggests that TC historians are preoccupied with the two key components of the Western progress mythology without serious consideration of how the larger culture came to choose Western scientific and technological styles.

Questions seldom asked

The significance of these contrasting levels of thematic discourse is further illuminated by considering six questions that are rarely addressed in the journal. Most have been noted in the analyses of chapters

2 through 4. Here our purpose is to reflect on the significance of their peripheral status in TC articles if the integration of design and ambience is indeed a central task for historians of technlogy.

1. "The roads not taken." The great majority of TC articles study technology by beginning with a technological event and looking backward to ask how this event came to be. It is true that detailed research into the origins of a new technology is an implicit critique of determinism. But from another point of view the absence of a complementary body of articles taking a technology as a starting point and "looking forward"—articles asking "what followed from the event?"—tends to create the impression that the history of technology is a set of success stories in which an event is the logical outcome of the historical forces identified in research. This is an elusive matter. Insofar as historians stress the fact that historical factors need not have come together as they did, they critique determinism. Insofar as they focus their attention only on those factors successfully contributing to the final outcome, they tend to foster determinism. TC's articles are overwhelmingly skewed toward studies of the origin of technological change and not to its results.

Only a handful of authors call for or provide historical studies of technological failures (appendix 3.21.1). Howard Mumford Jones, Louis Hunter, Robert Post, Eugene Ferguson, and Reinhard Rürup stress the danger of success stories—"company history"—as the typical mode of technological history. Their point is not that the study of successful technological achievements has no place in the field, but that the absence of a parallel literature studying the roads not taken in technological decision making creates the impression that, in Ferguson's words, "the whole history of technological development had followed an orderly or rational path, as though today's world was the precise goal toward which all decisions, made since the beginning of history, were consciously directed."[22]

Twenty-one historical articles discuss technological failures. They do not, however, all meet the criteria set down by Jones, Ferguson, and Rürup for avoiding the company history pitfall. Twelve take note of temporary lags within longer developments of ultimately successful technologies. Thus Robert Woodbury's debunking of the Eli Whitney myth refers to Whitney's failure to achieve interchangeable parts manufacture, but it is understood as part of the larger success of the technology.[23] Six articles treat failures resulting in the near disappearance of, or the notable noninvestment in, some design or tradition. Stuart Leslie, for example, describes the failure of General Motors to pursue its promising "copper-cooled" engine in the early 1920s due to a combination of internal jealousies and changing market perceptions by management.[24]

Although they do not discuss failures, three other articles contribute to this small body of discourse. Donald E. Thomas, Jr., writes of the deep ambivalence in Rudolph Diesel and his son Eugen concerning the "rationalist" ideology of technological progress at the turn of the twentieth century. John H. Jensen and Gerhard Rosegger call attention to the influence of colonial powers and, later, of Romanian nationalist engineers in creating the Fetesti-to-Constanta rail and bridge system. Their point is that the civil engineering project proved for a variety of reasons to be an expensive technical failure. Finally, Ruth Schwartz Cowan addresses the failure of Western technology to embody feminine modalities in its technological language or designs.[25] In their different ways, these nine articles provide us with historical instances of roads not taken in Western technological development. They counteract the autonomous technology myth precisely by calling attention to the fact that technological failures are not necessarily due to technical inferiority. In these cases other factors—cultural values such as nationalist pride, in-house corporate dissension, or dominant cultural value systems—were deciding factors.

As evidence of SHOT Whiggery it can be argued that these articles are both too few and too isolated from one another to constitute a significant contribution to the integration of design and ambience. None of them gives evidence of being part of a thematic community of discourse. As a group, they are also remarkably late in developing. With the exception of Esper (1965) and Kevles (1975) they appear in the final three years of the study.

2. Technology from the worker perspective. Apart from Samuel Lieberstein's study of the Soviet NOT movement, Richard Rosenbloom's analysis of nineteenth-century interpretations of worker tensions in mechanized society, Lewis Mumford's assertion that autocratic technics depend on slave labor, and Reinhard Rürup's critique that "very little has been done on the worker in history," we find only two types of references to workers in TC (appendix 3.21.2).[26] Eleven articles note some form of worker tension but only in passing. Daniel P. Jones is exceptionally explicit in identifying fear of labor strife as a primary motive for the introduction of tear gas in American police forces after 1921, but even this article fails to explore the nature of worker discontent in any significant detail.

Six articles treat cases of de-skilling, worker safety, or increased managerial surveillance with no reference to worker attitudes. Norman Wilkinson's study of Lammot Du Pont's Tayloristic efficiency studies is noteworthy in that his summary of questions raised by the case is innocent of any reference to what workers thought of such a movement toward radically increased managerial control.[27]

We see, therefore, nearly complete avoidance of the question of labor-management tension in the journal. The omission is one of the most substantial arguments for concluding that SHOT history has embodied the Whig perspective. If technological changes simply embody autonomous progress it would be consistent to treat workers as marginal and historically insignificant figures. If, on the other hand, technical designs embody cultural, economic, and political values, then the experience of workers who are intimately and sometimes violently affected by production system designs would seem to be an important perspective for interpreting these value-laden design changes.

3. Culture conflict in technology transfer. Only seven historical articles seriously consider cultural dimensions in technological transfers (appendix 3.21.3). This in itself is a significant lacuna in journal discourse, for transfer is a popular topic, the subject of seventy-six TC articles.[28] If technological designs embody the values of their culture of origin, cultural factors would be an expected dimension of transfer studies.

Another dimension of technology transfers, however, is even less well represented. Although the seven articles treat cultural tensions, the degree to which their authors stress the perspective of the recipient culture varies widely. When transfer is interpreted from the point of view of the receiving culture the question "Who wins and who loses because of the incoming technology?" is immediately raised. The question is subtle and demands careful understanding of the dynamics of the local culture. Transfer operates in a matrix of many different institutions and groups of people who, within the complex universe of their own world view, are variously affected by the value system embodied in the new technology. Such sensitivity to the subtleties of the recipient culture is normally lacking in TC transfer studies. Harold Dorn is more explicit than most. Writing of the Stalin regime's claim that Soviet industrialization was financed "without resort to oppressive loans from foreign governments nor yet to the conquest and spoliation of the weaker governments," he observes:

The remainder of the truth, however, which Vinter neglected to mention, is that in a backward, predominantly agricultural country the funds for rapid industrial development are obtained by squeezing the farming population between low prices for agricultural products and high prices for industrial goods.[29]

Even this very sepcific—by TC standards—reference to the receiving culture, however, is made in passing, without noting the bloody purges which, more than any other act, marked Stalin's squeeze of the peasants. Only Jensen and Rosegger, Bruce Sinclair, Shannon Brown, and Barton Hacker consider the resisters' point of view in any detail.[30]

It can be argued that the pervasive omission of the perspective of those injured by transfers fosters the autonomous Western technology myth. Were technology transfer simply part of the inevitable triumph of Western technology around the globe, then resisters would be of interest only because they "retard" the movement of progress. If, on the other hand, technology transfers are understood as culture-laden actions then these same resisters become interesting indeed. Their perceptions of the incoming technology would be an excellent locus for understanding the influence of cultural values in technical design.

TC is even more reticent about a related topic, Western colonialism. In a pattern strikingly similar to the avoidance of the workers' perspective, we find few studies where transfer is part of colonial exploitation and in those few the perspective of the colonized cultures is generally ignored (appendix 3.21.3). Thus while Mel Gorman notes in passing the hostility of the "conquered" to the construction of the first British telegraph in India, his central interest is the process of constructing the system from a British point of view.[31] Only Jensen and Rosegger address the question directly. Their portrayal of seventy-five years of competing Romanian transport systems is marked by a subtle shift of focus from the viewpoint of various colonial powers before 1881 to the perspective of the young native engineers who begin to emerge after Romanian independence. The recent concern with "appropriate technologies" in non-Western cultures in Hacker and DeWalt is the only other sign in TC of a serious consideration of the question.[32] The scanty representation of non-Western points of view in transfer studies suggests a chronic "addiction" for SHOT members to a Western world view.

4. Non-Western technologies. Historians of technology could also treat non-Western technologies as an object of study in their own right (appendix 3.21.4). A large group of such studies would foster comparisons among culturally distinct technological styles. Such comparisons would enhance the integration of technology and culture by revealing culturally relative characteristics in the designs themselves. Unfortunately, we see here again a substantial pro-Western bias. Of the 272 articles published in TC's first two decades only six treat non-Western technologies as their central consideration. Frankel's Indonesian pamor, Bray's Chinese plow, and Sleeswyk's Chinese clepsydra are all classic internalist treatments. Wachsmann and Kay's discussion of African musical instruments is more anthropological than historical and as such is untypical of the journal. Only two—E-Tu Zen Sun's biographical portrayal of the Chinese bureaucrat Wu Ch'i-chün and Billie DeWalt's study of the rural Mexican corn seeder—treat non-Western technologies from a contextual perspective. Other non-Western technologies appear in multiperiod treatments of single technologies such as Dresbeck's seven-thousand-year survey of the ski and Znachko-Iavorskii's similar study of cements. Once again we see a significant question that is not raised in TC discourse that might foster a design-ambient integration and a correlative critique of technological determinism.[33]

5. *Critiques of capitalism.* Not uncommonly critiques of capitalism are understood exclusively in a Marxist sense and, indeed, one serious consideration for us must be the complete absence of a Marxist perspective in TC's historical articles. Only seven articles explicitly consider a Marxist interpretation of technological change. Three, by Rürup, Rosenbloom, and Durbin, include it as one dimension of broad surveys of technological or worker studies. David Joravsky's work is exclusively devoted to a critique of extant Soviet history for avoiding its most significant issues. A. Zvorikine contributed two short articles articulating the current Soviet theory of technological history. Finally, Robert Heilbroner's "Do Machines Make History?" appears to adopt the Marxist premise of an inevitable movement from capitalism to socialism. No contextual case studies in TC adopt a Marxist perspective.[34]

The issue is broader than Marxism, however. Western capitalism, for good or ill, is a dominant factor in the cultural, economic, and political fabric from which Western technologies have emerged in recent centuries. It would be expected that a body of literature serious about integrating technology and culture would frequently address the influence of capitalism on technology. Here again TC reveals a striking lacuna.

Although interpretations of the innovation stage of emerging technology frequently refer to economic factors in capitalistic terms such as cost analysis and market forces, conscious examination of the relationship between capitalism and technical designs is rare (appendix

3.21.5). Otto Mayr's remarkable reflections on the link between feedback mechanisms and Adam Smith's economic theory is an exception. Besides Mayr, two articles call attention to ideological ambiguities in capitalism—Thomas's study of the Diesels and Dorn's treatment of Hugh Lincoln Cooper's relationship with Stalin.[35]

6. Women and technology. In the seventeenth volume of TC, Ruth Schwartz Cowan published the first article to consider women and technology. She has been joined recently by the pioneering members of WITH (Women in Technological History) in the attempt to interpret the relationship. These are positive developments, but by their scarcity and tardiness they only underscore the extraordinary bias in TC toward male figures and what are often seen as masculine values. The women's perspective is clearly as great an omission in journal offerings as the workers' perspective noted above.[36]

The problem is subtle and complex. Recent "gender and technology" literature has not arrived at a consensus about many pertinent questions (e.g., how to explain common cultural stereotypes about appropriate male and female behavior, to what extent and in what cultural circumstances women contributed to technological design and praxis, to what extent and how the modern Western "style" of technology reflects male-female inequities, etc.).[37] However these issues come to be resolved, women are as significant in the life of a culture as men. Their relationship with technological designs, whether as designers, users, or as a group who "lose" because of the way a technology has been designed, cannot be ignored in any fully articulated interpretation of the cultural ambience of technology. The assumption that existing technological designs are the result of an inevitable progress would, of course, justify relegating women to a marginal status in the story of technological change. The striking absence of the women's perspective therefore constitutes further evidence that SHOT remains trapped within the myth of progress.

The near-avoidance of these six questions is the most compelling evidence that SHOT historians have not escaped the perspective of Whig history. It would be reasonable to expect at least some of the issues to be vigorously pursued in a publication explicitly concerned with the integration of technology and culture.

It is important to note, however, that such omissions are not explicit arguments for the Whig assumption of autonomous progress. Historians are generally committed to conservative responsibility in their interpretations of historical events. To their credit, TC historians are nearly unanimous in their implicit, and sometimes explicit, repugnance for the autonomous progress myth. My brief reflections above on the openings toward a genuine contextualism do not really do justice to the *tone* of the journal's contextual articles. To anyone who has made a practice of reading these works over the years it should be obvious how radically different the language of these case studies is from the kinds of progress talk that oppressively din in our ears. (Forums such as advertisements for computers are perhaps the best contemporary locus for the myth in its pure form.) The care with which TC historians treat historical detail is in itself an affirmation of the specificity of each technology, of its rootedness in the everydayness of the human fabric.

Still, we should not take these six rarely considered questions too lightly. Scholars sin more by the questions they do not discuss than by irresponsibility in their treatment of the questions under consideration. This set of "sins of omission" constitutes the strongest argument that SHOT has failed in its purpose and has fostered the myth of autonomous technological progress despite the challenge it set for itself when it titled the journal *Technology* and *Culture*.

Recent Monographs and Integration of Design and Ambience

For all its virtues as a sample of discourse in the history of technology, TC is severely limited on one count. Articles are short, and since choices must be made among the many possible contextual factors, they cannot tell us how completely an author might integrate technology and cultural context in a longer work. It is helpful to supplement the the findings from TC with observations about some noteworthy historical monographs published in the field. What do SHOT's longer historical studies tell us about the society's achievements in creating a contextual style? Even with these studies the challenge is difficult because of the nature of technological history itself. The technological sophistication demanded by the discipline is so complex that it can absorb all of an author's attention when it comes to decisions about relevant evidence and the interpretative structure of the work. On the other hand, ambience is such a diffuse concept that it is very difficult to define it with enough focus so that it can function structurally on an equal basis with technological dynamics.

The problem SHOT historians have in achieving consensus about the meaning of ambience is highlighted when we recall who the society has defined as its disciplinary near neighbors. For various reasons the History of Science Society has been the overwhelming favorite in SHOT's first two decades. The popularity of the science–technology question in TC is one sign of its privileged status. Another is SHOT's consistent preference for joint annual meetings with the historians of science over any other group of scholars.[38]

Several other scholarly areas have been given some attention in the field. The importance of business and economic history can be seen especially in the area of emerging technology.[39] The newly created field of industrial archeology appears to have grown out of SHOT and continues to maintain close relationships with the society. Nonhistorical scholarship in the areas of the philosophy and sociology of technology has also made modest contributions. Finally, as will be noted below, the relationship between SHOT and labor historians has been surprisingly limited.

This pattern of interdisciplinary relationships suggests that the society has not established stable institutional relationships with any of its potential near neighbors except the historians of science. While science is indeed one aspect of the ambience of technology, it clearly cannot serve as a full definition of the concept. Given the diffuse character of SHOT's other scholarly ties, it is not surprising that the following survey of monographs shows no consistent single pattern, no consensus about how to define technology's ambience in its broadest sense.

This does not mean, however, that nothing has been achieved toward the toward the goal of a wide-ranging contextual history of technology. We will consider the question from two points of view. The first is a set of reflections on the sixteen books that have been awarded the Dexter Prize, SHOT's annual award for the best book in the field. The second is a survey of some contextual styles that have appeared in recent monographs within SHOT's ambience.

The Dexter Prize

The Dexter Prize-winning books (table 12) are hardly a complete set of monographs in the history of technology, but they do represent the society's judgment about the highest-quality research in each year since 1968. How have these seventeen books framed their study of technological history? Have any of them adopted a structural frame of reference in which technological sophistication is matched by some cultural or societal dimension, and have the books successfully integrated the several aspects?

Five books have maintained the traditional internalist perspective. Hans Eberhard Wulff's *The Traditional Crafts of Persia* (1968); Götz Quarg's translation of the fifteenth-century *Bellifortis*, a description of military art by Conrad Keyser (1969); Donald R. Hill's translation of *The Book of Knowledge of Ingenious Mechanical Devices*, a medieval account of Arabic machines and instruments (1974); Richard W. Bulliet's *The Camel and*

Table 12
Dexter Prize winners, 1968–1983

1968	Dr. Hans Eberhard Wulff, *The Traditional Crafts of Persia*
1969	Götz Quarg, ed., Translation and annotation of text and illustrations of Conrad Keyser's *Bellifortis*
1970	Lynn White, Jr., *Machina ex Deo: Essays in the Dynamism of Western Culture*
1971	Edwin T. Layton, Jr., *The Revolt of the Engineers: Social Responsibility and the American Engineering Profession*
1972	Thomas P. Hughes, *Elmer Sperry: Inventor and Engineer*
1973	Donald S. L. Cardwell, *From Watt to Clausius: The Rise of Thermodynamics in the Early Industrial Age*
1974	Daniel J. Boorstin, *The Americans: The Democratic Experience*
	Donald R. Hill, trs. and ed. Ism'il ibn al-Razzaz al-Jazari, *The Book of Knowledge of Ingenious Mechanical Devices*
1975	Bruce Sinclair, *Philadelphia's Philosopher Mechanics: A History of the Franklin Institute, 1824–1865*
1976	Hugh G. J. Aitken, *Syntony and Spark: The Origins of Radio*
1977	Richard W. Bulliet, *The Camel and the Wheel*
1978	Reese V. Jenkins, *Images and Enterprise: Technology and the American Photographic Industry, 1829 to 1925*
1979	David P. Billington, *Robert Maillart's Bridges*
1980	Louis Hunter, *Waterpower in the Century of the Steam Engine*
1981	David J. Jeremy, *Transatlantic Industrial Revolution: The Diffusion of Textile Technologies between Britain and America, 1790–1830*
1982	Edward W. Constant II, *The Origins of the Turbojet Revolution*
1983	Clayton R. Koppes, *JPL and the American Space Program: A History of the Jet Propulsion Laboratory*

the Wheel (1975); and David P. Billington's *Robert Maillart's Bridges* (1979) were all recognized for the care and depth of their treatment of technical questions. Their nearly one-third share of all Dexter awards is evidence that SHOT continues to see serious internalist scholarship as a major dimension of its work. They do not, however, provide us with any new interpretative models for the integration of design and the larger cultural ambience.

Lynn White, Jr.'s collection of essays, *Machina ex Deo* (1970) was recognized, no doubt in conjunction with his earlier *Medieval Technology and Social Change*, as a genuine innovation in the historiography of the field. The essays reflect White's awareness of the creative tension between Western technological style and the societal, philosophical, artistic, and particularly the religious world view of medieval Europe. As noted in chapter 1, White's work has long been acknowledged as a paradigm for the "new" approach to a contextual history of technology.

Edwin T. Layton, Jr.'s *The Revolt of the Engineers* (1971) is an exciting externalist work. It probes the peculiar nature of an engineering movement that was critical of capitalism while at the same time fostering the myth of progressive science and technology during the late nineteenth and early twentieth centuries. Although it does not provide an integration of technical design and cultural ambience, it explores the question by its detailed consideration of the philosophical views and the search for social status of American engineers in an era when engineering societies were making significant decisions about their role in the larger society.

In *Elmer Sperry* (1972), Thomas P. Hughes provides a design-ambient focus on the individual level by treating the personality and inventive style of Elmer Sperry as determinants in the designs of his inventions. By writing a richly textured biography, Hughes opens the question of inventive design to a larger world view where personal motives and problems can be seen as intimately involved in the design process.

The awards given to Donald S. L. Cardwell (*From Watt to Clausius*, 1973) and to Hugh G. J. Aitken (*Syntony and Spark*, 1976) demonstrate SHOT's continuing interest in the relationship between science and technology. Both authors, however, open their primary frame of reference to a broader context, Aitken by situating his account of the scientific and technological dimensions of early radio development in its economic context and Cardwell by attention to sociological and economic factors that influenced the prehistory and early history of thermodynamics.

The joint award for 1974, to Donald R. Hill's internalist translation and annotation of *The Book of Knowledge of Ingenious Mechanical Devices* and Daniel J. Boorstin's *The Americans: The Democratic Experience*, stands as a near paradigm of the methodological tension between internalist and externalist approaches. While Pulitzer Prize winner Boorstin is not generally considered a practicing historian of technology, his book was recognized by SHOT because of his use of a host of technological references for a new interpretation of the American national and cultural style. His importance for the field, giving legitimacy to the technological component of mainstream history, was duly noted in the award citation.

By adopting an institutional focus, Bruce Sinclair (*Philadelphia's Philosopher Mechanics*, 1975) and Clayton Koppes (*JPL and the American Space Program*, 1982) have been able to locate a remarkably wide range of technological and related scientific research projects in a richly textured historical and cultural context. In so doing, they give promise that institutional history may become a promising new contextual style for SHOT.

Reese V. Jenkins (*Images and Enterprise*, 1978), Louis Hunter (*Waterpower*, 1980), and David J. Jeremy (*Transatlantic Industrial Revolution*, 1981) all treat single technologies in great detail while situating them in a larger societal context. The three authors are alert to the significance of business and economic factors in the dynamics of complex technological change over a long time. In each book the relationship between technological and economic factors tends to overshadow any larger cultural frame of reference. This is particularly noteworthy in Jeremy's study of the transfer of British textile designs to New England, both in light of the scholarly attention that has recently been given to noneconomic aspects of the Lowell textile designs (especially in such works as Dublin's *Women at Work* and Kasson's *Civilizing the Machine*) and because of the observation, in chapter 4, of the importance of cultural factors in technology transfer.

Edward W. Constant II (*The Origins of the Turbojet Revolution*, 1982) has written a study of emerging technology that stands as a benchmark of the level of detailed sophistication achieved by SHOT historians in this thematic area. By focusing his attention on a single, well-defined moment in aviation history and by building on the work of earlier historical models of technological change and of the science–technology relationship, he has developed a complex new model that brings together intellectual, social, and institutional factors as they influence the emergence of a radically new technological design.

What can we conclude from this overview of Dexter Prize-winning books? Like TC articles, they treat a wide range of technologies and introduce many different contextual factors. It is clear, too, that the two most popular themes found in the articles are well represented. Cardwell, Sinclair, Aitken, Jenkins, Hunter, Jeremy and especially Hughes and Constant have created complex interpretations of emerging technology. Cardwell, Sinclair, Aitken, Jenkins, and Constant also address the science–technology relationship.

More than the articles, however, a number of the contextual books have taken important steps toward integrating technical design and the larger human context. SHOT's alertness to this challenge is evident in the language of the citations for the four most recent awards. After discussing Louis Hunter's command of waterpower technology the citation concludes: "Above all, Hunter has produced a study that integrates technology with economic and social factors without submerging it."[40] David Jeremy's citation states: "What makes the work especially valuable is its integration of technical, economic, and social perspectives in depicting the nature of industrialization in early American textile manufacturing."[41] Constant's *Turbojet* is praised for the same commitment to design-ambient integration: "Worthy of special note are his well-argued comparison of national technological traditions and his emphasis on the importance of ideology and even a degree of fanaticism among inventors who commit themselves to radically new directions."[42] What is striking in these citations is the use of such expressions as "above all," "especially valuable," and "worthy of special note," a newly consistent emphasis by the awards committee on the integration of design and context.

Most recently and perhaps more tellingly, the awards committee cites Koppes's contextual achievement in *JPL* without any language suggesting that it is extraordinary.

> In the process of discussing the major technical achievements of JPL, Koppes keeps the reader aware of *the larger context* in which these achievements took place. He examines the role of contract research, government and university bureaucracy, industrial relations, personal ambitions, national and international politics that provide the setting.[43] (my italics)

By 1983 the subtleties of Koppes's contextualism, including his treatment of the specifically cultural question of "the conflict between ideology and policy of the national security state and traditional values of a democracy" are not treated as exceptional, except for Koppes's quality in articulating them.

The difficulty of the integration has not escaped notice. Arnold Thackray's review of Sinclair's Franklin Institute study compliments the author for his attention to a multifaceted array of factors (especially the institute's commitment to the advancement of knowledge, its internal and sociological history, and its situation as an urban Philadelphia institution) but suggests that he falls somewhat short of a thorough integration. "Professor Sinclair possesses a thoughtful awareness of all these things. However, he is sometimes uncertain which of the three foregoing modes of analysis [noted above] he desires as primary, or whether he seeks an integration."[44] Thackray's observation highlights the problem indicated at the beginning of this section. Do we find in SHOT's book-length studies the emergence of a new frame of reference in which technological and cultural aspects operate in an equal and integrated fashion to interpret the design-ambient relationship? Sinclair's work is seen to be a clear, though limited step in this direction.

The continuing presence of classic internalist works in the list as late as 1979 indicates that contextual history has not become an overwhelming favorite for the society. The years in which internalist works are cited either represent the absence of contextual work of equal quality or they represent diversity in the awards committee over time about what kind of scholarship best represents the society. In either case they remind us again of the slow and difficult process of creating a radically new paradigm for the history of technology, the many-sided character of the task, and the extent of the historical terrain that remains underdeveloped.

Contextualist trends in recent monographs

Community studies. Two recent community studies, Merritt Roe Smith's *Harpers Ferry Armory* (1977) and Anthony F. C. Wallace's *Rockdale* (1978), have attracted considerable attention in SHOT circles. This is partly due, no doubt, to their role as twin signals of increasing legitimation for the history of technology in the Organization of American Historians and in anthropological circles. The recognition by Wallace, a leading cognitive anthropologist, of the importance of technological detail in Rockdale's development was matched by the Organization of American Historians' recognition of Smith, a historian of technology, as winner of the Frederick Jackson Turner Prize. The nearly simultaneous publication of their books was taken as indicative of a breakthrough in the scholarly acceptance of technology in social history and anthropology.

In his commentary on the two books, Glenn Porter characterizes both works as significant breaks from SHOT traditions.

Not only do these studies consider technical change, which in and of itself is not especially of interest to the historical world outside S.H.O.T., but they consider technology's impact on the human beings who employed and operated it, and they analyze its connections with the history of business, religion, politics, popular culture, and so on. They do not represent still more studies of great men, they do not worry about "firsts" in the history of technology, and they are concerned *much more* with the *adoption* and impact of new technology than with their *invention*. In all these respects I think they are models for historians of technology.[45]

Porter's comment corroborates a major premise of this chapter. Given TC's propensity for studies of the origins of technological change, particularly in its thematic treatment of emerging technology, it would seem that these analyses of the adoption and impact of technological designs begin to provide a necessary balance. We have pointed out that a body of discourse that overly stresses the "success story" motif inherent in the study of most emerging technology interpretations tends to foster the ideology of autonomous technological progress. Careful contextual studies of technological successes are freed from the progressive ideology when they are situated within a larger interpretative frame that includes just such questions as Smith and Wallace address. Smith in particular succeeds not only in integrating the design of machine tool technology with the community culture of Harpers Ferry but also in studying typical emerging technology themes such as invention, development, and innovation within a larger context that includes the issues of philosophical, economic, political, and social tensions related to its adoption.[46]

The size of the community is central to the books' value as methodological exemplars. This limited and manageable frame of reference, "the community," serves as a major structural component of each study. Within that frame of reference both authors have managed to study an exceptionally rich array of cultural factors as they change over time. A second equally complex and coherent structural frame of reference is the dynamic of changing technological systems. When the two are treated as historiographical equals they provide a compelling example of how human culture shapes technical design and how already-designed technologies influence later cultural developments. Thus it would appear that the recognition given the two works in SHOT circles is due not only to their role in legitimating the field in larger scholarly communities, but also to the creative contribution they make to the contextual challenge that SHOT has set for itself from its earliest days.[47]

Worker perspective. As we have indicated, the history of workers has been largely ignored in TC articles. This is only one sign of SHOT's curious avoidance of a worker perspective. In contrast to the strong tradition of joint annual meetings with the History of Science Society, historians of technology have shown little interest in formal relationships with the Labor History Society.[48] Nevertheless, recent developments among both historians of technology and labor historians suggests that a change is under way. Members of both scholarly traditions are beginning to recognize significant mutual interests.

Hugh J. G. Aitken's *Taylorism at the Watertown Arsenal* (1960) is an early, rare, and excellent contribution to the worker–technology relationship. Although it is primarily a study of management, it portrays worker response to the imposition of Taylorist practices in rich detail.[49] It is only in recent years, however, that such serious consideration of workers has begun to catch on in SHOT circles. Merritt Roe Smith's *Harpers Ferry Armory* is a bellwether for this new interest. His account of the introduction of interchangeable parts manufacturing gives the world view of armory workers as much prominence as that either of the little community's aristrocratic "junto" or of the enforcing Ordnance Department officers. David Noble, who gained notice in the field with his study of American corporate strategy and philosophy in *America by Design* (1979), has shifted his focus in the direction of a worker perspective in his *Forces of Production*, a study of numerically controlled machine tools. Beginning in 1976, a handful of papers read at SHOT annual meetings have focused on workers.[50]

A similar movement has begun to appear in several important monographs by labor historians. David Brody's ground-breaking *Steelworkers in America: The Non-Union Era* (1960) has been followed recently by Thomas Dublin's *Women at Work: The Transformation of Work and Community in Lowell, Mass., 1826–1860* (1979) and Stephen Meyer's *The Five Dollar Day: Labor, Management and Social Control in the Ford Motor Company, 1908–1919* (1981). All three authors have chosen to center their interpretation of labor-management relations around detailed analysis of the designs of production technologies.

The importance of these complementary movements should not be underestimated. The critical historiographical value of this combination of technological detail with labor-management tensions lies in the visibility it gives to the role of a management world view and value system in the design of production technology. When such design changes are treated without consideration of what workers thought or felt about them and without attention to worker resistance, the impression can be given that the changes occurred for reasons of efficiency and that the commitment to efficiency operated in abstraction from any larger cultural world view on the part of management. As soon as the workers are taken into serious consideration, however, it becomes evident that two radically distinct world views often meet, in conflict, at the point of technological design itself. These conflicts manifest the cultural relativity inherent in technology because the design in question embodies one and not both world views. By so demonstrating, these studies elegantly situate the process of technological change within a complex and believable historical ambience.

Labor historian Herbert Gutman has created one of the most helpful thematic models to date for interpreting the cultural significance of these tensions. By identifying the conflict between the "culture" of America's immigrant workers and the "society" of the management-designed work place, he introduces the notion of culture conflict into the design process itself.[51]

Calling attention to this convergence of labor and technological studies is not intended as a suggestion that labor history is the only important perspective for historians of technology. It is clear, however, that avoidance of a worker perspective in treatments of production technology is a serious omission, a subtle reinforcement of that ideology of technological progress which claims that all design innovations represent the "march of progress" in the culturally superior technological West. From this perspective the emergence of technological history that gives serious consideration to workers is a long overdue and very positive development.

Institutional history. Bruce Sinclair's history of the Franklin Institute, Edwin Layton's *Revolt of the Engineers*, Reese Jenkins's *Images and Enterprise*, and Alfred Chandler's *The Visible Hand* demonstrate the value of institutional history in two ways. Sinclair and Layton contribute to the contextual approach by stressing the complex relationships between research and professional engineering institutions and the larger culture within which they flourished. Jenkins and Chandler embody the same approach, at least in part, by integrating technical design with economic institutions. While culture clearly encompasses more than economic factors, their sophistication in linking the domains of business and technological history is a positive contribution.[52]

Another development that promises to expand the scope of institutional history is the emergence of a group of scholars probing the relationships between the U.S. government (particularly military and aerospace institutions, with their attendant values and styles) and technological design. The spate of serious monographs studying the U.S. aerospace programs has produced one Dexter Prize winner—Clayton R. Koppes's *JPL and the American Space Program*—and some significant probing of the role of institutional matrixes in technological change. Few are more explicit than Alex Roland's *Model Research: The National Advisory Committee for Aeronautics, 1915–1958* (1984), David Kite Allison's *New Eye for the Navy: The Origins of Radar at the Naval Research Laboratory* (1981), and Pamela E. Mack's recent dissertation, "The Politics of Technological Change: A History of Landsat" (1983). The clearest sign of an emerging thematic focus on the military-technology relationship is the forthcoming collection, *Military Enterprise and Technological Change: Perspectives on the American Experience*, edited by Merritt Roe Smith.[53]

Systems Approach. The discussion of the systems approach in chapter 2 stressed the common tendency to see systems as hierarchical and standardized. Edward Constant's study of the turbojet revolution and especially Thomas Hughes's recent comparative study of German, British, and American light and power systems (*Networks of Power*, 1983) take the study of technological systems and emerging technology to a new level of awareness of the cultural dimensions of technical design. Both authors reveal their sensitivity to the contextual question by their attention to the external forces that influence the design and the maintenance of the systems they study. Hughes in particular is noteworthy on two counts. He explicitly moves beyond a mechanistic interpretation of systems by adopting the distinction between open and closed systems: "An open system is one that is subject to influences from the environment; a closed system is its own sweet beast, and the final state can be predicted from the initial condition and the internal dynamic."[54] By choosing to study electrical power networks under both rubrics he begins to move beyond the limitations of the hierarchical and standardized model. As a further step in this direction, he situates his systems approach in a compelling example of comparative studies in technological styles, an often praised but seldom practiced virtue.

Gender and technology. The recent appearance of *Dynamos and Virgins Revisited* (1979), edited by Martha Moore Trescott, and *Machina ex Dea* (1983), edited by Joan Rothschild, and *Scientific-Technological Change and* *the Role of Women in Development* (1982), edited by Pamela M. D'Onofrio-Flores and Sheila M. Pfafflin, together with Ruth Schwartz Cowan's *More Work for Mother* (1983), are indications that the question of gender and technology no longer languishes in a historiographical backwater.[55]

In this varied array of recent books we find signs of the increasing commitment among historians of technology to an integration of design and context on a level that does justice to social history and to an understanding of culture as a human artifact itself. While these trends do not yet constitute an overwhelming consensus among SHOT scholars, they are an impressive sign of vitality, a strong suggestion that the commitment to contextual history of technology has recently deepened and matured.

The rest of this chapter will propose a new interpretative model that may take the field a step further in this process. Some of its elements come from the patterns already seen in TC articles and in the monographs just considered. The model is a first attempt, on my part, at seriously confronting the problem of a single perspective for integrating design and ambience. It is hoped that the model will give due weight to the many complexities of such a perspective, but of course, as a first attempt, it calls for critique and revision by historians of technology.

Three Constituencies in Technological Change

The following model rests on the assumption that attention to design constraints and ambient factors *together* creates a valid contextual approach. Every contextual historian decides how much technical detail to include in the written story s/he tells. Too much detail bores the reader and blurs the lines of larger historical interpretation because of its distracting complexity. Too little detail renders the interpretation unintelligible by not giving enough insight into the way specific design constraints interact with other historical factors. However much or little technical detail the historian chooses to include, however, s/he must have achieved enough intimacy with the design so that the entire interpretation, design constraints and ambient factors alike, is intelligently treated. Technical practitioners lament that interpretations of technological artifacts commonly reveal woeful ignorance of the difficulties and subtleties facing those who design or maintain them.

On the other hand, the historian of technology must possess an equal familiarity with the historical fabric of the time and place in which the story is situated. It is clearly as great a failing to write history in ignorance of historical complexities as it is to write of technology in ignorance of design constraints.

This design-ambient definition of the history of technology assumes that every technology is local-specific. At every stage of its development, from invention to obsolescence, a technology can be explained historically only in terms of relationships with its particular ambience. The value of this definition can be further understood if we contrast it with the systems approach to emerging technology. Both approaches are applicable to many levels of technological complexity. They differ in that the systems approach concentrates on internal or functional constraints, while the design-ambient model focuses on the encounter between internal design and surrounding ambience. We could say that the systems model stresses *design* integration alone as the intelligible core for understanding technology, while the design-ambient model stresses negotiation—between integrated design and the tangle of exogenous ambient realities—as the intelligible core for that understanding.

The following model identifies three stages of development in any successful technology and three related constituencies. Although they will be presented serially, it will be obvious that many of the dynamics are at work simultaneously.[56]

Three stages of a successful technology: design, momentum, and senility

The attractiveness of the autonomous technology interpretation derives from two striking ways that Western technological systems in the past several centuries have influenced the future. In the first place, they have fostered further technological changes in a linear fashion. The machine tool system of standardized manufacturing, for example, has been the locus for a century and a half of advances in cutting tool metallurgy, in tool design itself, and in production system integrations. At the same time, these systems have clearly compelled societal adaptations to their technical style. The standardized manufacturing system has also fostered a host of conforming responses by society in such areas as labor-management relations, consumer habits, and so forth. These dynamics must be accounted for in any historical interpretation of technology. However, the argument here is that they are not historically intelligible without first attending to a prior moment in the history of every technical design, a moment when the design is not yet a source either of further technical change or of societal change. This moment can be called the "design stage."

The design stage and the design constituency. The design stage is characterized by the flexibility of the technology in question. For every successful technology there existed a period when neither its later adoption nor its specific design constraints were inevitable or obvious. The principle pertains more precisely to breakthrough than to marginal innovations. Any successful technology will continue to generate design changes within its paradigm over much of its lifetime. Insofar as these marginal innovations are governed by the technical frame of reference from which they emerge, they tend to reveal fewer of the characteristics of flexibility and noninevitability noted here. They tend to be predictable and, within their technical frame of reference, more or less "inevitable." Care must be taken, however, not to overemphasize the distinction between marginal and breakthrough innovations because the relationship between predictable and unpredictable novelty is never completely clear. Every new design has some elements of flexibility about it even when it is seemingly inevitable.[57]

The flexibility of the design stage means that no technological design is fixed in advance. For the historian who seeks out such moments of origin it is critically important to tell the story in a way that does justice to both design and ambient factors. The source of any original design is an individual or group who can be called *the design constituency*, the first of the three constituencies to be found in this model. To interpret how this constituency arrived at a given technological design the historian must first consider what technical constraints influenced their work. Thus the group at Ford Motor Company who created the moving assembly line that is characteristic of "Fordism" worked out of the technical tradition of machine tool production systems that had developed through the nineteenth and early twentieth centuries in America.[58] It would be historically frivolous to argue that the creators of Fordism had total flexibility when they began their project after 1908.

On the other hand, the Ford design constituency was not simply a mechanical component of the inevitable unfolding of the mass-production system. Careful historical interpretation reveals many nontechnical influences at work on the designers. For example, Ford and the people working with him appear to have shared a set of values related to an ideal of American life, sometimes called the "melting pot," in which immigrants were seen as backward, uncivilized, and potentially chaotic; immigrant workers need paternalistic help so that they could "rise" to America's level of civilization and rigorous control lest they disrupt the American social fabric. Again, the Ford constituency shared the common value system of American management that viewed the shaping of corporate policy and of the structures of the work place as the job of a governing group of managers rather than as the task of all participants negotiating with one another.[59]

A careful history of this design stage would take into account not only the technical constraints under which the design constituency operated but also the larger world view and value system of the group. The ambient side of the design-ambient relationship would be given primacy precisely because the flexibility of the design stage demands a nontechnical explanation of why things turned out as they did. Thorough history would ask, "Who were the members of the design constituency, and what were their motivating values?" "How did they understand reality, and how did that world view influence their interaction with existing technological constraints?"

The momentum stage and the maintenance constituency. The flexibility characteristic of a time of origin gradually moves toward design rigidity. This is due in part to the nature of the design process itself, in part to the psychological dynamics of those who design, and in part to the way a successful design influences its ambience. In the design process, each decision about how the components of the design will relate to one another precludes alternative relationships. With each decision, more options are foreclosed as a single pattern takes shape. Once a design is complete it is often true that seemingly minor changes would involve so many other relational changes as to be infeasible. In other words, the more mature the design the more further innovations within it will be governed by the structural lines already set down. The design constraint side of the design-ambient tension assumes greater and greater influence in such marginal innovations.

Changes in an already determined design are difficult too because of the increasingly vested interests of the design constituency as it goes through the difficult work of the design process. To one who has spent great energy in solving vexing problems, suggestions that would demand radical change tend to be unwelcome. Even less welcome are suggestions that the entire project be abandoned, as in the case when the original reason for the project disappears.[60]

These two movements from flexibility toward rigidity explain some of the momentum achieved by an increasingly successful technical design. The momentum of a technology is defined as the combination of its growing design rigidity and the increasing difficulty of changing or stopping its activity. The momentum is further enhanced by the patterns of societal adaptation to a given design. As a design becomes invested in, not only within the personal motivation of its designers but also in the economic, political, and social habits of its ambience of adoption, the design exerts still further influence on that ambience. In our example it is clear that the Ford style of mass-produced automobiles exerted a continually greater influence on the whole structure of American society for decades. More and more people came to depend on the auto companies for employment, and the automotive influence spread to such automobile-related businesses as gas stations, motels, insurance companies, and drive-in stores. America's political system adapted to the automobile, not only legislating licensing and driving speed, but also by committing public funds to highway construction. Finally, the life-style and value systems of Americans grew to reflect automotive mobility, with particular emphasis on individualism in transportation styles. The more American society adapted to the mass-produced automobile, the greater the momentum of the original Ford design.[61]

For the historian, interpreting the momentum stage requires a different set of questions than does the design stage. Because of the growing rigidity and power of the technological design, the historian's attention tends to shift from the nontechnical motives of the design constituency to the technologically structured motives of what can be called *the maintenance constituency*. This constituency includes all the individuals, groups, and institutions that have come to depend on the design and consequently have adapted to its constraints. Because they both profit from and depend upon it they maintain its momentum in society and become a primary source of its power to affect future technological and societal directions. Thus the historian would be concerned with questions such as: "Who are the people and the institutions who profit from the technology?" "What strategies do they adopt to foster its momentum?" "How do they preclude developments that would threaten the design's continued momentum?" "What further technological developments are promoted by their commitment to the technology?" And finally, "How is their world view—personal and societal—affected by their investment in and dependence on the technology in question?"

By studying such matters the historian addresses precisely those dynamics that are normally adduced as evidence for the existence of an autonomous technological progress. In the momentum stage the technical design itself exerts the most powerful influence on further developments. Were the momentum stage the only aspect of the technological dynamic to be considered it might be possible to construct an interpretation where the only role of nontechnical factors was to foster and adapt to technology. Instead of talking about "a" technology, as one necessarily does when referring to the flexible design stage, the language of progress begins to talk about "technology" as an independent sociotechnical force. Historical studies of technologies in the momentum stage are, therefore, dependent on the perspective gained by studies of the design stage if they are to remain free of the progress ideology.

The senility stage. Every technology is a human artifact, an artificial construction whose design reflects a limited set of prior technical constraints and a limited set of values within a particular world view. It is inevitable that the technology's life cycle will eventually lead to a form of "senility." The term is chosen because of the nature of human senility—the increasing disconnection from one's ambience due to a hardening of the arteries. The metaphor calls attention to the liabilities of human rigidity during the aging process. A technology can be considered senile when inherent rigidities cause relationships with its ambience to become dysfunctional. The larger ambience in which every technology exists is, of course, much more complex than the technology itself. Precisely because the larger world is not systemic, because it is continually changing under the influence of a host of exogenous variables that resist inclusion in any humanly constructed system, that larger world will not retain its "fit" with any successful technology forever. When new ambient forces—coming either from the emergence of radically new technologies, from shifting value commitments, from an increasing resistance of the ecological order to pressures coming from the technology's modes of operation, or from any other source—meet in such a way that they render a technology's normal operation unacceptable, that technology has reached a stage of senility.

It is possible, of course, that the maintenance constituency of such a technology will find the way back to the radical flexibility of a new design stage, that it will restructure its technology so thoroughly that a new synthesis of design and ambience emerges. More commonly, however, due to the multiple rigidities already noted, the maintainers of the technology will attempt to force the ambience back into configuration with the old design. One might call these strategies "fascist" in the sense that an increasing amount of violent force is exerted on the ambience in order to preserve the viability of the rigid system.

For the historian who chooses to study a technology in its senility stage, the questions of interest revolve around the responses of the maintenance constituency members. "How do they view the new ambient forces that are putting pressure on the normal functioning of their technology?" "How open are they to radical change within the technology over which they preside or on which they depend?" "What strategies do they take to preserve or to transform the technology?" By calling attention to such questions the historian underscores the governing insight of the contextual style of history, which rests on the premise that "technology" as a historical dynamic is made up of many specific designs that are neither inevitable nor eternal. Each is a cultural artifact rooted in the limitations of its time and place of origin, of flourishing, and of senility.

The impact constituency. We have now seen all three stages of the dynamic of a successful technology and two of its constituencies. A third constituency remains, one often ignored in treatments of technology despite its ability to reveal the historical limitations of every technical design. *The impact constituency* can be defined as the set of individuals, groups, or institutions who lose because of the design of a technology, those who suffer from the rigidities and limitations inherent in the technical dynamic.

Impact constituencies are, almost by definition, difficult to locate historically. It is not simply that the powerless leave few records for the historian to ponder, although this is certainly true. It is also because it requires a radical act, on the part of the designers or maintainers of a technology, to give visibility to those who lose due to their technology of preference. It is even true that those who lose tend, in the face of powerful technologically based momentum, to remain unaware of how they lose because of the system's design.[62] Progress talk is often the barrier that leaves members of the impact constituency in the shadows of history. It is often argued that those who "temporarily" seem to suffer from the advance of progress will eventually gain because of that advance, or at least that subsequent generations will do so. If it is true that successful technological designs are successful precisely because they embody the march of human progress, then the claim that all will eventually benefit holds.[63] If, on the other hand, it is true that every technological design reflects a limited set of values and technical styles, and if those limitations render every technology culture-bound and finite, then the world view, the values, and the technological preferences of those who lose become critically important for the historian of technology. They provide a perspective that, more than any other, reveals the humanness and limitation of every technological artifact, and in so doing they form one of the most helpful ways of escaping the myth of progress in historical interpretation.

The impact constituency is an elusive reality on several counts. Neither the design constituency nor the maintenance constituency tends to perceive its viewpoint, and its members are sometimes unconscious of their loss as well. This situation is complicated further when we realize that members of both the design and maintenance constituencies can also be members of the impact constituency. It is a commonplace observation that every technology has "second order consequences," results that were not anticipated or desired during the design process. Thus someone who helped to design a technology might well find, as the system achieves momentum, that some of its effects turn out to be personally harmful. Even more obvious is the possibility for someone to both profit from and lose because of a system's momentum. Automobile assembly-line workers, for example, often speak of the dehumanizing quality of line work even though they are attached to their jobs as a livelihood.

To see the impact constituency in its purest form, it is necessary to shift one's attention to those marginal places where people feel the impact of a system "from the outside," where their own preferred world order is invaded by the technology. Studies of technology transfer, especially from one cultural tradition to another, provide perhaps the clearest examples of impact constituencies. It is only under the rubric of the ideology of progress that the sufferings, the anger, and often the despair of small cultures invaded by dominant technologies can be dismissed as historically irrelevant. Every culture constitutes a whole world view, giving meaning to the entire human endeavor, technological and nontechnological alike. When such a culture is forced to accept a technology that embodies a world view contradicting its own wisdom, the historian can identify in great detail the violent impact of the invading technology.[64]

Nevertheless, the contextual perspective in the history of technology can alert the historian to the existence of the impact constituency within the culture of origin as well. By continually asking the related questions "Who designed the technology?" "What values did they embody in the design?" "Who gains what from the design?" and finally, "Who loses what because of the design?" the historian situates the design in its full historical context and makes a complete break from the ideology of progress.

Technological styles

One final nuance remains. For the sake of simplicity I have limited the discussion of this model to individual technologies. It should be clear, however, that the approach can be extended to a more synthetic interpretation of technological styles. A technological style can be defined as a set of congruent technologies that become "normal" (accepted as ordinary and at the same time as normative) within a given culture. They are congruent in the sense that all of them embody the same set of overarching values within their various technical domains. For example, it can be argued that the United States, beginning with the U.S. Ordnance Department's 1816 commitment to the philosophical ideal of standardization and interchangeability, gradually adopted a set of normal technologies that incorporate that ideal. From this point of view many distinct technological developments—the machine tool tradition, the growth of standardized and centrally controlled rail systems, the centralization and standardization of corporate research and development, the use of consumer advertising to program individual buying habits, the increasing centralization and complexity of electricity and communication networks, etc.—can be interpreted as participating in a single style, embodying a specific set of values within a specific world view. In short, the dynamics of interaction between world view and technical design need not be limited to an individual technology.

On this level of technological style the contradictions between the ideology of progress and a contextual interpretation become most starkly clear. For the first ideology this sketch of twentieth-century "normal technologies" in the United States would portray the inevitable advance of progress. From the contextual approach, the entire set of normal technologies would be subject to the common historical pattern I have described. After a period of initial flexibility during which the key design decisions were made, and a period of momentum during which the set of technologies flourished, the contextual historian would expect to see signs of senility, when the inherent limitations and rigidities of the style would begin to be dysfunctional because of changes within the larger contextual ambience. From such a perspective, the awesome power of standardized and standardizing technologies in the United States would lose their aura of inevitability and eternity and take their place in the family of human technological styles. Like all human artifacts, they would reveal their nobility, their violence, and their mortality as the tale of human history unfolds.[65]

The three-constituency model as the locus for thematic questions

It is clear from this discussion that no historian of technology can hope to write a full history of any technology in less than a multivolume effort. By proposing the constituency model it is not my intent to deny the validity of more limited frames of reference. The wide variety of more specific approaches to history identified in this book can all find their place within the model. Studies of emerging technology, of the nature of technological cognition, of technology transfer and technological momentum, and the array of contextualist patterns in recent monographs—community studies, worker history, institutional history, the new systems approach, and the perspective of women—are all congruent with the constituency model.

What is at stake here is the larger frame of reference within which historians situate their more limited studies. It is the argument of this entire book that historians of technology must choose between the ideology of progress and the design-ambient contextual approach as the governing model for their research. Both perspectives provide an encompassing model within which more specific studies can be situated, but the differences between the two are radical. It is impossible to adhere to the ideology of progress and write contextual history; the two views are incompatible to the core.

The purpose of articulating the three-constituency model is to challenge the pervasive Western tendency to assume autonomous progress. While the details of the model can surely be critized and refined, it is my premise that historians of technology will recognize the design-ambient approach to history as the underlying basis for their commitment to a genuine contextualism.

By its nature contextual history is a vulnerable process in which the historian is deeply affected by the humanity of the subject matter. To reject as ahistorical the ideology of autonomous progress is to recognize that technological designs are intimately woven into the human tapestry and that all of the actors in the drama, including the storyteller, are affected by tensions between design and ambience. By telling the stories of technological developments while respecting the full humanity of the tale, the contextual scholar rescues technology from the abstractions of progress talk and, in the process, takes part in the very ancient and very contemporary calling of the historian, reweaving the human fabric.

Appendix 1 Taxonomies Used in This Study

The following two appendixes present the taxonomies that are the basis for the analysis of TC's articles in chapters 2 through 5. This appendix introduces the following taxonomies by describing the procedure I used in constructing them, discussing the taxonomic theory that has been helpful to me, and finally giving a brief description of each taxonomy.

The procedure leading to the final taxonomies was basically a process of reading and rereading the 272 articles while attempting to identify the most helpful analytical models for interpreting all of the articles as a single body of shared historical discourse. Whenever a new model suggested itself I attempted to construct an initial taxonomy embodying the model while doing justice to the linguistic usage of the articles. The test of each model was always the same: was it possible to score the individual articles on a set of subcategories constituting the various dimensions of the model? Frequently, of course, the attempt to score an article would lead to an insight about TC's usage that demanded a change in the design of the taxonomy. Occasionally, too, a taxonomy that had been created and scored was abandoned because it did not reflect actual usage or because the information it presented was not judged to be helpful for interpreting TC's usage. Such a process is, by its very nature, always open to further revision. The final taxonomies are the result of six years of creation and construction. It is my hope that they will prove helpful to the reader, as they have to me, in understanding all of TC's articles as a single body of historical discourse.

The purpose of a taxonomy is to create a frame of reference in which a number of discrete discussions within a community of shared discourse can be organized into a meaningful whole. The taxonomies here fall into two basic categories, which I have titled "exclusive" and "inclusive." An exclusive taxonomy creates a set of subcategories so designed that any article can score in one and only one logical space within it. The taxonomy is exclusive because scoring an article in one space excludes, in principle, scoring it in any other. The exclusive taxonomy is also designed to include every article in the set.[1]

An inclusive taxonomy creates a set of subcategories designed to reflect the asymmetrical process in which historians create a language of thematic interpretation. The subcategories reflect the contribution of various articles to the creation of a complex and multifaceted concept. No two articles will express the concept in the same fashion. Thus the design of the taxonomy must make logical spaces available for every subcategory that has been seen as a significant dimension of the concept by a number of authors. For example, if the thematic concept called "the process of emerging technology" is found to contain three distinct subcategories (invention, development, and innovation), we may find articles that refer to one, two, or all three of the subcategories. Some articles may not score anywhere on the taxonomy because they do not treat the process of emerging technology at all. Such a taxonomy is called inclusive because it is designed to include all article references to such subcategories within it.[2]

The validity of any taxonomy is often said to depend on two criteria. First, the taxonomy must accurately embody the conceptual model being presented. Every intellectually distinct dimension of the model must be represented by a specific logical space within it.[3] Second, the scoring of the articles on the taxonomy must be replicable; that is, other scorers must be able to replicate the scores if they are provided with the operational definitions of each subcategory.[4] This criterion is meant to reduce the problem of subjectivity in the scoring process.

When a taxonomy is used to interpret historical discourse, however, the constraints of strict replicability create a problem. To construct a taxonomy that permits strictly replicable scoring we must break the general concept into quantifiable and univocal units that can be identified and counted by any independent scorer. For the complex concepts that occur in historical thematic discourse, a perfectly replicable taxonomy would entail the creation of an extraordinary number of discrete subcategories. Even if such a set of subcategories were economically feasible, however, the fact that they must be discrete and univocal would not allow them to reflect the overlapping and nonquantitative nature of the language with which historians interpret themes. Thus a strict adherence to the canon of replicability tends to result in a restriction of taxonomies to readily quantifiable data such as monetary figures or to data that is trivial because it oversimplifies historical discourse. As a result, I have chosen to modify the criterion of replicability in the following manner.

In this study the validity of a taxonomy depends on its heuristic helpfulness to the members of the scholarly community, i.e., SHOT, who participate in the shared historical discourse. If the structure of a taxonomy and the interpretation of individual articles based on it prove helpful and enlightening for scholars in the field, if the taxonomy reveals methodological or thematic dimensions of the entire group of articles which foster a deeper awareness of presuppositions operative in the community, and if it results in a reading of the articles that rings true to scholars who have themselves read them, then the taxonomy is heuristically valid.[5]

The taxonomy in appendix 2, "Three Dimensions of Methodology," is the only exclusive taxonomy used. It has been designed to score every TC article in terms of three methodological dimensions containing sixteen permutations. The taxonomies presented in appendix 3 deal with the very complex thematic language of the articles. Their purpose is twofold: to help the reader visualize the entire theme in a single frame of reference, and to help locate those articles contributing to the language of each theme's subcategories.

Notes

1. For several discussions of what I am calling an "exclusive taxonomy," and for the theory of taxonomic analysis generally, see the following studies of cognitive anthropology. Wallace, "Culture and Cognition," pp. 116–118; Bruner, Goodnow, and Austin, "Categories and Cognition," pp. 183–184. See also Holsti, *Content Analysis*, p. 99.

2. For discussions of the asymmetrical character of inclusive taxonomies and their contrasts with exclusive taxonomies, see Wallace, "Culture and Cognition," pp. 118–120; Bruner et al., "Categories and Cognition," p. 185.

3. On the expression "logical space" see Wallace, "Culture and Cognition," p. 117. On the criterion that a valid taxonomy accurately embodies its conceptual model, see Holsti, *Content Analysis*, p. 95.

4. On the canon of replicability see Holsti, ibid.

5. Holsti is well aware of the subjectivity of even the most strictly replicable of taxonomies. "Many of the most rigorously quantitative studies use non-numerical procedures at various stages in the research. This is likely to be the case *in initial selection of categories*" (ibid., p. 11; my italics).

Appendix 2 Three Dimensions of Methodology

Although the operational definitions for each subcategory in the three dimensions of methodology are discussed in Chapter 1, it may help to present them in a format designed to show the entire taxonomy in overview. This appendix will begin with the operational definitions of all three dimensions. Tables 13 and 14 present the shifting proportions of each dimension over the years of TC's publication. Finally, a complete list of the articles clustered according to the sixteen logical spaces of the taxonomy is included.

Operational Definitions

1. Methodological style

Contextual style. The article discusses the functional design of the given artifact(s) and also discusses some aspect(s) of the ambience in which the artifact(s) exist(s).

Internalist style. The article focuses only on the functional design of the given artifact(s).

Externalist style. The article discusses some technological ambience without discussing the functional design of any artifact(s) that may pertain.

Nonhistorical essay. The article does not adopt a historical perspective.

Historiographical essay. The article discusses problems involved in historical research about technology.

2. The function of hypotheses in argumentation

A priori. The article bases its argumentation on one or more explicitly stated hypotheses that have been articulated prior to the research on which the article's findings are based.

A posterior. The article articulates new hypotheses based on the evidence uncovered during its own research.

3. Organizational format

Case study. The article frames its subject matter as a single event, within the pertinent specific time and place constraints.

Expanded study. The article creates an enlarged frame of reference either by including several types of technology, in one or more time periods, or by tracing a single technology through a series of discrete events and time periods.

Technology and Culture
Articles Scored on Three Dimensions of Methodology

1. Contextual / A priori / Case study

Bailes, Brittain, Brown, Bryant (22, 24), Cardwell (36), Chapin, Constant, DeWalt, Emmerson, Esper, Feller, Fox, Fries, Garcia-Diego, Hacker (89, 91), Harrison, Hewlett (103, 104), Hoberman, Hounshell, Howard, Hughes (112), Jensen and Rosegger, Jeremy, Layton (144), Mayr (156), Multhauf (174), Pearson, Post (191), Reynolds, Roland, Scherer, T. Smith, TeBrake, Uselding, Vincenti, Welsh (259), White (260), Wise, Woodbury (42 articles)

2. Contextual / A posteriori / Case study

Boyer, Bryant (21), Burke (27), Cain (32, 33), Cardwell (37), Carver, Clark, Condit (43, 44), Dalrymple (49), Dorn, Dornberger, Frazier, Fullmer, Gorman, Hagen, R. Hall, Hills and Pacey, Hughes (111), R. Hunter, Jenkins, Jewett (122), Krammer, Leslie, Loria, Massouh, Mayo-Wells, Mayr (155), McCutcheon, Miles, Multhauf (171, 172), Parr, Perry, Poole and Reed, Pursell (196), Rasmussen, Sandler, Sharrer, Sinclair (225), Skramstad, M. Smith, Tascher, Tokaty, Tucker, Vogel, Wik (263, 264), Wilkinson (266) (52 articles)

3. Contextual / A priori / Expanded study

Balmer, Burns, Cowan (48), Ferguson (71), A. Hall (93, 94), Hanieski, Hughes (113), Jevons, Kerker, Layton (143), Mumford (176), Price (193), Rosenberg (211, 213), Rosenbloom, Ruttan and Hayami, Schallenberg and Ault, Woodruff and Woodruff (21 articles)

4. Contextual / A posteriori / Expanded study

Bedini (10), Bilstein, Bryant (23), Burke (27), Dalrymple (50), Flick, Gade, Glick, Hilliard, L. Hunter, Kahn, Kraus, Leighton (146), Miller, Nunis, Fulton, and McCarthy, Paterson, Rosenberg (212), Sinclair (226), C. Smith (232, 233), Welsh (258), White (261), Wilkinson (265) (21 articles)

Table 13
Three dimensions of methodology

	Case study		Expanded study		Totals
	A priori	*A posteriori*	*A priori*	*A posteriori*	
Contextual	42 (15%)	52 (19%)	21 (8%)	21 (8%)	136 (50%)
Internalist	9 (3%)	5 (2%)	22 (8%)	11 (4%)	47 (17%)
Externalist	7 (3%)	15 (6%)	8 (3%)	7 (3%)	37 (14%)
Nonhistorical	31 (11%)	1 (4%)			32 (12%)
Historiographical	11 (6%)	9 (4%)			20 (7%)
Totals	100 (37%)	82 (30%)	51 (19%)	39 (14%)	272 (100%)

Table 14
Changes in the proportion of the five methodological styles over time

	1959–1966	1967–1973	1974–1980	Totals
Contextual	45 (41%)	38 (53%)	53 (59%)	136 (50%)
Internalist	20 (18%)	13 (18%)	14 (16%)	47 (17%)
Externalist	10 (9%)	13 (18%)	14 (16%)	37 (14%)
Nonhistorical	24 (22%)	6 (8%)	2 (2%)	32 (12%)
Historiographical	11 (10%)	2 (3%)	7 (8%)	20 (7%)
Totals	110	72	90	272 (100%)

5. Internalist / A priori / Case study
Bachrach (4), de Camp, Fryer and Marshall, Kilgour, Mark, Abel, and Chiu, Reti (204, 205), Simms, Sleeswyk (9 articles)

6. Internalist / A posteriori / Case study
von Braun, Edelstein, Jewett (121), Packer, Roach (5 articles)

7. Internalist / A priori / Expanded study
Bachrach (5), Bowles, Boyer, Bray, Davison, Frankel, Fussell, Hartenberg and Schmidt, Kanefsky and Robey, Kohlmeyer and Herum, Kreutz, Leicester, Lienhard, Muendel, Nelson, Puhvel, Rae (201), Schmandt-Besserat, Shelby, C. Smith (230, 231), Usher (22 articles)

8. Internalist / A posteriori / Expanded study
Bedini (9), Condit (45), Dresbeck, Hacker (90), Kren, Leighton (147), Pendray, Spence, Spong, Strauss, Znachko-Iavorskii (11 articles)

9. Externalist / A priori / Case study
Harris and Pris, Layton (141), Post (192), Reingold, Robinson, Tobey, Virginski (7 articles)

10. Externalist / A posteriori / Case study
Brittain and McMath, Claxton, Keller (129), Kevles (131, 132), LaForce, Lieberstein, Multhauf (170), Overfield, Pursell (197, 198), Rezneck, Sun, Susskind and Inouye, Thomas (15 articles)

11. Externalist / A priori / Expanded study
Calder, Cowan (47), Drucker (58), Durbin, Heizer, Multhauf (169), Nicholas, Shriver (8 articles)

12. Externalist / A posteriori / Expanded study
Bachrach (3), Bedini (11), Ferguson (69), Finn, Mazlish, Pursell (199), Vanek (7 articles)

13. Nonhistorical / A priori
Agassi, Baranson, Buchanan, Bunge, Clarke, Drucker (59), Dryden, Ellul, Feibleman (66, 67), Finch, Goldschmidt, Hartner, Heilbroner, Howland, McLuhan and Nevitt, Meeker, Mesthene (162, 163, 164), Morris, Mumford (177, 178), Price (194), Skolimowski, Solo, Strassman, Theobald, Wachsmann and Kay, Watson-Watt, Zvorikine (272) (31 articles)

14. Nonhistorical / A posteriori
Gilfillan (1 article)

15. Historiographical / A priori
Burlingame (30), Daniels, Drucker (57), Ferguson (70), Geise, Jones, Layton (142), Mayr (157), Mumford (175), Rae (200), Zvorikine (271) (11 articles)

16. Historiographical / A posteriori
Allen, Burlingame (29), Hughes (114), Joravsky, Keller (128), Kranzberg, Multhauf (173), Rürup, White (261) (9 articles)

Appendix 3 Major Themes in Technology and Culture

This appendix has been designed to help the reader follow the complex discussion of the many subordinate themes discussed in chapters 2 through 5. It depends on the complete bibliography, with short titles, of TC's 272 articles following the appendixes. The subordinate themes are presented here in the same sequence as in the text. Every article is identified by author and when necessary by its number in the bibliography. The appendix can also be used as a detailed thematic index of the book because each set of articles is accompanied by page references to the discussion of the subtheme.

The process of scoring TC's articles, as described in appendix 1, is endless in principle. It is certainly conceivable that articles could be found which belong on one of the following lists and which have escaped my attention. My claim is only that the articles listed refer to the theme or subordinate theme in question, not that the lists are necessarily exhaustive.

Emerging Technology (Chapter 2)

1. The act of invention

1. **Verification of the true inventor or the origin of an invention (p. 40)**
   Balmer, Bedini (9), Bowles, Bray, Brittain (18), Bryant (21, 22, 24), Cardwell (37), Chapin, Dresbeck, Emmerson, Fox, Fryer and Marshall, Hacker (90), Hounshell, H. Jones, Keller (128), Kerker, Kreutz, Leighton (146), Mayr (156), Puhvel, Post (192), Reti (204, 205), Robinson, Roland, Scherer, Schmandt-Besserat, C. Smith (233), Tucker, White (260, 261), Wik (264) (35 articles)
2. **Communication of the significance of the invention (pp. 40–41)**
   Brittain (18), Bryant (22, 24), Chapin, DeWalt, Hounshell, Post (191, 192), Reynolds, Robinson, Wise (11 articles)
3. **The nature of the act of insight (pp. 41–42)**
   Agassi, Constant, Hughes (113), White (261) (4 articles)

4. **The personality and motivation of the inventor (pp. 42–43)**
Brittain (18), Constant, Ferguson (71), Fryer and Marshall, Harrison, Hounshell, Jensen and Rosegger, Mayr (156), Robinson, Scherer, C. Smith (233), Watson-Watt, Wise (13 articles)

5. **The intellectual background of the inventor (p. 43)**
Brittain (18), Bryant (21, 22, 24), DeWalt, Fox, Fryer and Marshall, Hounshell, Jenkins, Kerker, Layton (144), Mayr (156), Reynolds, Scherer (14 articles)

6. **Inventive insight as origin of innovation process (pp. 43–44)**
Bryant (21, 22, 23, 24), Chapin, Constant, DeWalt, Finch, Hewlett (104), Jenkins, Mayr (156), Rürup, Scherer (13 articles)

7. **Inventive insight as the conceptual element throughout the process of invention (pp. 44–45)**
Agassi, Brittain (18), Bryant (23, 24), Hounshell, Jevons, Kerker, Layton (142), Robinson (9 articles)

2. Development

1. **Developmental activity based on model building and testing of models (pp. 45–46)**
Brittain (18), Bryant (21, 24), Chapin, Constant, Dornberger, Frazier, Hagen, R. Hall, Hanieski, Harris and Pris, Hewlett (103, 104), Hills and Pacey, Hughes (112, 113, 114), Jenkins, Jewett (122), Kraus, Layton (144), Leslie, Mayr (156), Miles, Nunis, Fulton, and McCarthy, Pearson, Pendray, Perry, Post (191), Rasmussen, Reynolds, Roach, Robinson, Ruttan and Hayami, Sandler, Scherer, T. Smith, Solo, Strassmann, Tascher, Tokaty, Vincenti, Virginski (43 articles)

2. **Need to lobby for funding for development project (p. 46)**
Dornberger, Flick, R. Hall, Pearson, Perry, Post (191), Sandler, T. Smith, Vincenti (9 articles)

3. **Military setting for development project (p. 46)**
Dornberger, Frazier, R. Hall, Hewlett (103, 104), D. Jones, Krammer, Miles, Pearson, Perkins, Perry, Sandler, M. Smith, Tokaty (14 articles)

4. **Corporate setting for development project (p. 46)**
Brittain (18), Bryant (21, 24), Chapin, Hills and Pacey, Howard, Hughes (112, 113), Jenkins, Jeremy, Leslie, Mayr (156), Robinson, Scherer, Wise (15 articles)

5. **Development as a goal-directed activity (pp. 46–49)**
Bryant (21, 24), Chapin, Constant, Dornberger, Frazier, R. Hall, Hewlett (103, 104), Howard, Hughes (112, 113), Jenkins, Jeremy, Leslie, Mayr (156), Miles, Perry, Post (191), Robinson, Ruttan and Hayami, Scherer, T. Smith, Tokaty, Wise (25 articles)

6. **Nonthematic references to problems solved during development (p. 47)**
DeWalt, Dorn, Hagan, Harris and Pris, Jensen and Rosegger, Jewett (122), D. Jones, Kraus, Nunis, Fulton and McCarthy, Pendray, Rasmussen, Roach, Tascher, Tucker, Virginski (15 articles)

3. Innovation

1. **Contrasts between innovation and invention (pp. 51–53)**
   Bryant (24), Constant, Hughes (114), Kohlmeyer and Herum, Layton (142, 143), Mayr (156), Robinson, Rürup, Scherer, Uselding (11 articles)

2. **Innovation as characterized by profit motive or market strategy (pp. 52–54)**
   A. Cost analysis, quantity production, standardization
   Clark, DeWalt, Dornberger, Feller, Harrison, Howard, Hughes (111, 113), Jenkins, Jensen and Rosegger, Jeremy, Jevons, Krammer, Layton (144), Massouh, Paterson, Post (191), Robinson, Rosenberg (211, 213), Scherer, Sinclair (225), Strassman, Uselding, Wik (264), Woodbury (26 articles)
   B. Corporate setting
   Bilstein, Bryant (21, 24), Chapin, Clark, Constant, Feller, Howard, Hughes (111, 113), Jenkins, Jeremy, Jevons, Jewett (122), Kohlmeyer and Herum, Krammer, Leslie, Loria, Massouh, Mayr (156), Multhauf (171), Paterson, Perkins, Robinson, Rosenberg (211, 213), Scherer, Sinclair (225), Strassman, Uselding, Wik (264), Woodbury (32 articles)
   C. Patent strategy
   Bryant (21), Chapin, Robinson, Wise (4 articles)

3. **Innovation as diffusion process (pp. 54–55)**
   Bilstein, Bryant (24), Burke (28), Clark, Constant, DeWalt, Feller, D. Jones, Harrison, Hughes (111), Kanefsky and Robey, Kohlmeyer and Herum, Krammer, Perkins, Reynolds, Rosenberg (211), Ruttan and Hayami, Sinclair (225), M. Smith, Tucker (20 articles)

4. **Nonanalytic verification of the diffusion of a specific technology (p. 55)**
   Bowles, Bray, Cowan (47), Fox, Kreutz, Muendel, Vanke, Virginski, Wik (264) (9 articles)

4. Technological Support Network

1. **Support network as a necessary condition for emerging technology (pp. 61–62)**
   Baranson, Bryant (24), Cardwell (37), Condit (43), Dalrymple (49), Daniels, Dornberger, Drucker (58), Finch, A. Hall, (93, 94), Hanieski, Heilbroner, Howard, Hughes (111, 113), Jewett (121), Kanefsky and Robey, Kerker, Kohlmeyer and Herum, Lieberstein, Miller, Multhauf (172), Post (191), Price (194), Pursell (196), Rae (200), Reynolds, Rosenberg (211), C. Smith (231), T. Smith, Spence, Tucker, Uselding, Usher (35 articles)

2. **Infrastructure as generating new technology (pp. 62–63)**
   Drucker (58), Krammer, Price (194), Rosenberg (213), Usher (5 articles)

3. **Infrastructure as related to emerging technology (no causal assertion) (p. 63)**
   Bilstein, Cain (33), Cardwell (36), Hewlett (103), Jeremy, Jewett (121), LaForce, Paterson, Ruttan and Hayami, Scherer, Sinclair (225), Solo, Vincenti, Welsh (258), Wik (264) (15 articles)

4. **Impact of new materials or techniques on support network (p. 63)**
   Dalrymple (49), Dornberger, Jewett (122), Leighton (146), Reynolds, Rosenberg (213), Skramstad, Uselding (8 articles)

5. **Impact of new artifact on support network (pp. 63–64)**
Bilstein, Cain (33), R. Hunter, Massouh, McCutcheon, Sinclair (226), Wilkinson (265) (7 articles)

5. Technical Tradition

1. **Explicit definition of technical tradition (p. 64)**
Constant, Hughes (114), Uselding, White (261) (4 articles)
2. **Tradition fostering incremental invention (p. 65)**
Bray, Condit (45), DeWalt, Dryden, Feibleman (66), Frazier, Hacker (90), R. Hall, Howard, Jensen and Rosegger, Pearson, Price (193), Rae (201), Reynolds, Rosenberg (213), Sandler, Skramstad, Vincenti (18 articles)
3. **Tradition as background for breakthrough inventions (pp. 65–66)**
Bryant (22), Condit (44), Constant, R. Hall, Hanieski, Jenkins, Kerker, Post (191), Uselding, Welsh (259), Wik (264), Woodbury (12 articles)
4. **Social mechanisms for transmission of knowledge in a tradition (pp. 66–67)**
Bryant (21, 23), Cowan (48), DeWalt, Howard, Hughes (113), Mayo-Wells, Multhauf (172), Vincenti, Wik (264) (10 articles)
5. **Tradition as cumulative knowledge (p. 67)**
Dalrymple (50), Mumford (175), Vincenti (3 articles)
6. **Tradition as hindrance to invention (p. 67)**
Feibleman (66), Hilliard (2 articles)
7. **Verification of the influence of prior tradition on a specific invention (p. 67)**
Bedini (9, 10), Bowles, Bray, Kren, Reti (204, 205), C. Smith (232), White (260, 261) (10 articles)
8. **Description of events in a tradition without analysis (p. 68)**
Balmer, von Braun, Garcia-Diego, Hacker (89), Hoberman, Kilgour, Kraus, Mayr (155), Parr, Reti (205), Tascher (11 articles)
9. **Breakthrough inventions generating subsequent incremental inventions (p. 68)**
Constant, R. Hall, Lienhard, Kerker (4 articles)
10. **Other references to impact on subsequent tradition (p. 68)**
Condit (44), Layton (144), Leslie, Pearson, Pendray, Vogel, White (260) (7 articles)

6. Systems

1. **Explicit theory of systems (pp. 69–72)**
Constant, Drucker (57), Hughes (112, 113), Mayr (155), Rosenberg (213), Strassman (7 articles)
2. **Single machine as system (p. 72)**
Bryant (21, 23), Chapin, Constant, Feibleman (67), Hacker (89), Hanieski, Howard, Kraus, Leslie, Post (191), Scherer, Vincenti, White (261), Wik (264) (15 articles)
3. **Development project as system (pp. 72–73)**
von Braun, Bryant (24), Dornberger, Hagen, R. Hall, Howard, Hughes (113), Leslie, Mayo-Wells, Miles, Perry, Rasmussen, T. Smith (13 articles)

A. Development case studies using "system" explicitly
R. Hall (p. 424), Miles (p. 482), Perry (p. 469) (3 articles)

4. **Production unit as system (pp. 73–75)**
Chapin, Fries, Harris and Pris, Hills and Pacey, Jenkins, Jeremy, Pursell (196), Robinson, Scherer, Sinclair (225), M. Smith, Uselding, Wilkinson (266) (13 articles)

5. **Transmission network as system (pp. 75–76)**
Agassi, Balmer, Bilstein, Brittain (18), Cain (32), Clark, Hounshell, Hughes (111, 112, 113), Jensen and Rosegger, Jewett (122), Massouh, Sinclair (226), Skramstad (15 articles)

6. **References to systems as standardized (pp. 75–76)**
Chapin, Clark, Drucker (57), Feller, Fries, A. Hall (94), Hills and Pacey, Jenkins, Jeremy, Kraus, Miles, Rasmussen, Rosenberg (211), Ruttan and Hayami, Sinclair (225), M. Smith, Strassmann, Uselding, Usher, Wilkinson (266) (20 articles)

7. **Sector of economy as system (pp. 76–78)**
Carver, Dalrymple (50), Gade, A. Hall (94), Hilliard, Kohlmeyer and Herum, Lieberstein, Rasmussen, Rosenberg (211, 213), Ruttan and Hayami, Schallenberg and Ault, Usher (13 articles)

8. **Impact of emerging technology on a system (p. 79)**
Burke (27), Rosenbloom, Rürup, Schallenberg and Ault (4 articles)

The Science vs. Technology Relationship (Chapter 3)

7. Statement 1: Scientific activity is motivated by curiosity, whereas technology is motivated by the desire to solve problems (pp. 86–87).

1. **Hypothesis as limited conclusion from a case study**
Dornberger, Hewlett (103), Scherer, T. Smith, Wise (5 articles)

2. **Hypothesis as a universal assumption**
Buchanan, Bunge, Mumford (177), Rae (200, 201), Skolimowski (6 articles)

3. **Explicit critique of the hypothesis**
Hughes (112), Meeker, Reingold, C. Smith (231) (4 articles)

8. Statement 2: The "desired artifact" in science is a theoretical model, whereas knowledge is in the service of the "desired artifact" of technology (pp. 87–90).

1. **Adoption of the hypothesis**
Agassi, Bryant (21, 23), Buchanan, Cardwell (36), Constant, Hacker (89), Hewlett (103, 104), Hughes (112), Layton (141, 142, 143, 144), Multhauf (171), Overfield, Price (194), Pursell (197), Rae (201), Reingold, Reynolds, Skolimowski, C. Smith (231, 232, 233), Vincenti (26 articles)

2. **Technological knowledge as theoretical**
Agassi, Bryant (21, 23), Constant, Hughes (112), Jewett (121, 122), Layton (141, 142, 143), Overfield, Price (194), Pursell (197), Reingold, Skolimowski (15 articles)

3. **Technological knowledge as empirical**
Hacker (89), Multhauf (171), C. Smith (231, 232, 233) (5 articles)

9. Statement 3: Science fosters technological creativity and rationalizes existing technological practice (pp. 90–91)

1. **Case studies: science as necessary condition for inventive insight**
Brittain (18), Constant, Fox, Hewlett (103, 104), Kerker, Reynolds, Scherer, T. Smith (9 articles)
2. **Case studies: science governs development processes**
Bryant (23, 24), Hughes (112), Loria, Mayr (156), Multhauf (172), Paterson, Rosenberg (211, 212) (articles)
3. **Case studies: science rationalizes existing technological practice**
Condit (43, 45), Multhauf (171, 172, 174), Overfield, Rae (201), Rezneck, Tobey (9 articles)
4. **General theory**
Bunge, Feibleman (66, 67), Gilfillan, A. Hall (94), Kevles (132), Kohlmeyer and Herum, Leicester, Rürup, Watson-Watt, Zvorikine (271, 272) (12 articles)

10. Statement 4: Technology contributes to science by creating instruments, by posing scientific problems, and by creating conceptual models for later science (pp. 92–93)

1. **Technology creates instruments for science**
Bedini (10), von Braun, Burlingame (30), Dryden, Feibleman (66), Hagen, Mayo-Wells (7 articles)
2. **Technology poses problems for science**
Bryant (21, 24), Bunge, Feibleman (66), Finch, Kerker, Layton (141), Rae (200, 201), C. Smith (231, 232) (11 articles)
3. **Technology creates conceptual models for science**
Buchanan, Cardwell (36), Fryer and Marshall, Kerker, Mayr (155) (5 articles)

11. Statement 5: Scientific and technological activities that occur in human communities often influence science-technology interactions (pp. 93–94)

1. **Communities with clearly defined boundaries**
Cardwell (37), Kerker, Layton (141, 142), Price (194) (5 articles)
2. **Communities with unclear boundaries**
Fullmer, Jensen and Rosegger, Kevles (132), Multhauf (169, 174), Overfield, Post (192), Pursell (197), Reingold, Reynolds, Robinson (in discussion of Hughes [112]), Wise (12 references)

12. Statement 6: Technology is applied science (pp. 96–99)

1. **Use of the expression "applied science"**
Bunge, Feibleman (66, 67), Heilbroner, Kevles (132), Leicester, Morris, Pursell (199), Rae (200), Watson-Watt (10 articles)

2. **Historical transition from "craft" to "exact science"**
   Bunge, Condit (43), A. Hall (94), Lienhard, Leicester, Heilbroner, Rürup (7 articles)
3. **Science as the sole source of knowledge in modern technology**
   Bunge, Feibleman (66), A. Hall (94), Leicester, Rae (200, 201), Rürup, Zvorikine (271, 272) (9 articles)

13. Statement 7: Technology is not applied science (pp. 99–103)

1. **Explicit critiques of the applied science hypothesis**
   Cardwell (36), Drucker (58), Hughes (112, 114), Jevons, Layton (141, 142, 143), Multhauf (174), Skolimowski, T. Smith, Thomas (12 articles)

Characteristics of Technological Knowledge (Chapter 3)

14. Scientific concepts

1. **Scientific concepts appropriated by technology and restructured according to its purposes (pp. 103–105).**
   Agassi, Bryant (23, 24), Fullmer, Hagen, Hewlett (104), Hoberman, Hughes (112), Kerker, Layton (141, 142, 143, 144), Mayr (156), Multhauf (174), Perkins, Perry, Price (194), Reynolds, Skolimowski, T. Smith, Strassman, Tobey, Usher, Wise (25 articles)

15. Problematic data

1. **Need for data in cases of emerging technology (p. 106)**
   Brittain (18), Bryant (23, 24), Clark, Cardwell (37), Dalrymple (50), Hacker (89), Hewlett (103, 104), Hoberman, Hounshell, Hughes (112), Jewett (122), Layton (144), Leslie, Mark, Abel, and Chiu, Mayr (156), Multhauf (172), Paterson, Perkins, Reynolds, Sinclair (115), Tascher, Vincenti (24 articles)
2. **Need for data due to problems occurring in normal use (pp. 106–107)**
   Burke (27), Cain (32, 33), Flick, Fullmer, Hills and Pacey, Jeremy, Layton (141), Mark, Abel, and Chiu, Multhauf (171, 172, 174), Overfield, Parr, Paterson, Reingold, Rezneck, Rosenberg (212), Sinclair (225), Skramstad, C. Smith (231, 232, 233), Znachko-Iavorskii (24 articles)
3. **Data search institutionalized in testing facilities (p. 107)**
   Burke (27), Hanieski, Jeremy, Kevles (132), Leslie, Paterson, Pursell (197), Reingold, Rezneck, Sinclair (225), Vincenti (11 articles)

16. Engineering Theory

1. **General description: formal experimental theory about artifacts (pp. 107–109)**
   Balmer, Brittain (18), Brittain and McMath, Bryant (23, 24), Cardwell (37), Condit (43, 44, 45), Constant, Finch, Garcia-Diego, Hacker (89), Hanieski, Hughes (112, 113), R. Hunter, Jewett (122), Kevles (131), Kohlmeyer and Herum, Layton (141, 142, 143, 144), Lieberstein, Mayr (156), Multhauf (172), Post (191), Reingold, Robinson, Rürup, Sinclair (225), Skolimowski, C. Smith (231), T. Smith, Tucker, Vincenti (37 articles)

2. **Engineering theory in cases of emerging technology (pp. 109–110)**
Brittain (18), Bryant (24), Constant, Hanieski, Harrison, Hughes (112, 113), Layton (144), Mayr (156), Multhauf (172), Post (191), Reynolds, Robinson, Sinclair (225), T. Smith, Tucker, Vincenti (17 articles)

3. **Institutional base for engineering theory (pp. 110–113)**
Brittain and McMath, Hughes (112, 113), Kevles (132), Kohlmeyer and Herum, Layton (141, 142), Lieberstein, Mayr (156), Multhauf (172), Price (194), Pursell (198), Reingold, Sinclair (226), T. Smith, Vincenti (16 articles)

4. **Engineering theory contrasted with technical skill (pp. 113–114)**
Brittain and McMath, Bryant (23), Cardwell (37), Condit (44), Dorn, Finch, Garcia-Diego, A. Hall (93), R. Hunter, Jewett (122), Multhauf (172), Post (191), Sinclair (225) (13 articles)

17. Technical skill

1. **Skill as learned experientially (pp. 113–116)**
   A. Skill as intimacy
   Feibleman (67), C. Smith (233) (2 articles)
   B. General descriptions of skill
   Balmer, Brittain and McMath, Bryant (23), Cain (33), Claxton, DeWalt, Dorn, Dornberger, Feibleman (66, 67), Finch, Garcia-Diego, A. Hall (93), Hewlett (104), Hughes (114), Jensen and Rosegger, Jeremy, Layton (141), Leslie, Lieberstein, Massouh, Mayr (156), Morris, Paterson, Reynolds, Rosenberg (211), Rürup, C. Smith (231, 233), Vincenti (30 articles)

2. **Skilled labor (pp. 116–118)**
   A. Needed in a technical process
   Baranson, Boyer, Brittain and McMath, Brown, Burns, Dornberger, Feibleman (67), Finch, Garcia-Diego, Glick, Howard, Jensen and Rosegger, LaForce, Lieberstein, Pursell (198), Rosenberg, (211, 213), Simms, M. Smith, Solo, Strassman, Welsh (258), Wilkinson (265) (23 articles)
   B. Replaced by machines or standardized processes
   Clark, Cowan (47), Esper, Feller, Fries, Harris and Pris, Jenkins, Jeremy, Packer, Pursell (196), Rasmussen, Rosenbloom, Sinclair (225), M. Smith, Welsh (259), Wilkinson (266) (16 articles)
   C. Attention to labor-management tension over replacement of skilled labor by machines
   Brown, DeWalt, Harris and Pris, D. Jones, Krammer, Pursell (196), Rosenbloom, Thomas (8 articles)

3. **Skill generates nontheoretical rules for praxis (pp. 118–119)**
Brittain and McMath, Bryant (23, 24), Burns, Cain (33), Dornberger, Edelstein, Feibleman (66, 67), Finch, Garcia-Diego, A. Hall (93, 94), Hewlett (104), Hills and Pacey, R. Hunter, Jeremy, Layton (143), Massouh, Multhauf (172, 174), Packer, Parr, Post (191), Rosenberg (211), Rürup, Sharrer, Sinclair (225), C. Smith (231), Uselding, Welsh (258) (31 articles)

4. **Skill codified in simple mathematical formulas (p. 119)**
Burns, Drucker (57), Glick, Hacker (89), Kahn, Layton (141), Mayr (155), Shelby, Simms (9 articles)

Technology and Its Cultural Ambience (Chapter 4)

18. Technology transfer

1. **Verification of specific transfers (pp. 123–124)**
   - A. Verification as the central focus
     Bachrach (3), Fox, Roland, Tokaty, Virginski, Wilkinson (265) (6 articles)
   - B. Verification included but not as central focus
     Balmer, Bedini (10), Bowles, Burke (28), Dalrymple (50), Ferguson (71), Garcia-Diego, Harris and Pris, Jewett (121), Kren, Kreutz, Muendel, Multhauf (171), Puhvel, Sleeswyk, White (260, 261), Woodbury (18 articles)
2. **Vehicles of technology transfer (p. 124)**
   - A. Theoretical models of transfer
     Feller, Rosenberg (211), Ruttan and Hayami, Woodruff and Woodruff (4 articles)
   - B. Skilled personnel as vehicle
     Brown, Cain (33), Daniels, Drucker (59), Feller, Gorman, R. Hunter, Layton (141), Loria, Rezneck, M. Smith, Virginski, Wilkinson (265) (13 articles)
   - C. Journals, exhibitions, schools as vehicles
     Carver, Ferguson (69), Frazier, Rezneck, Roland, Sinclair (225), Skramstad, Woodbury (8 articles)
   - D. Formal agreements as vehicle
     Dalrymple (49), Dorn, Fries, Hacker (91), Hughes (111), Krammer, Kraus, LaForce, Loria, Pursell (196) (10 articles)
   - E. Colonial policy as vehicle
     Gade, Gorman, Hoberman, Jensen and Rosegger, Leighton (146), McCutcheon, Woodruff and Woodruff (7 articles)
   - F. Espionage as vehicle
     Jeremy, Wilkinson (265) (2 articles)
3. **Transfer and technological support network (pp. 124–128)**
   - A. Theoretical models
     Baranson, Calder, Drucker (59), Goldschmidt, Rosenberg (211), Ruttan and Hayami, Shriver, Theobald, Woodruff and Woodruff (9 articles)
   - B. Case studies focused on transfer
     Brown, Dalrymple (49), Dorn, Feller, Fries, Hacker (91), Hoberman, Hughes (111), Jensen and Rosegger, LaForce, McCutcheon, Nicholas, Pursell (196), Wilkinson (265) (14 articles)
   - C. Peripheral references
     Brittain and McMath, Carver, Daniels, DeWalt, Gade, Gorman, R. Hunter, Jeremy, Jewett (121), Kraus, Leighton (146), Loria, Mayr (156), Rosenberg (211, 212, 213), Rürup, Sinclair (225, 226), Skramstad, C. Smith (232), Tokaty (22 articles)
4. **Transfer and Culture (pp. 128–134)**
   - A. Technology as culturally neutral
     Goldschmidt, LaForce, Morris, Rürup (4 articles)
   - B. Prescriptive articles calling for respect of cultural values
     Baranson, Calder, Drucker (59), Shriver, Theobald (5 articles)

C. Case studies focused on transfer and culture
Brown, Dorn, Fries, Jensen and Rosegger, Hacker (91), Hughes (111), Sinclair (226) (7 articles)

D. Peripheral references to transfer and culture
Bilstein, Brittain and McMath, Dalrymple (49), DeWalt, Gorman, D. Jones, Hughes (114), LaForce, Pursell (196), Rosenberg (212), Ruttan and Hayami, Welsh (258) (12 articles)

19. Technological determinism

1. Split of efficiency norm from cultural norms (pp. 136–139)

A. Critiques
Buchanan, Clarke, Ellul, Ferguson (70), Hartner, Meeker, Morris, Mumford (175), Pursell (199) (9 articles)

B. Example of the position
Feibleman (67), Lienhard, Usher (3 articles)

2. Fixed sequence of technological progress (pp. 140–143)

A. Critiques
Durbin, Ferguson (70), Joravsky, Hughes (114), Rürup, Shriver, Thomas, White (261) (8 articles)

B. Example of the position
Burlingame (30), Heilbroner, Lienhard, Mesthene (162, 163), Price (193, 194), Watson-Watt, Zvorikine (271, 272) (10 articles)

3. Society must adapt to technological change (pp. 143–145)

A. Critiques
Ellul, Daniels, Pursell (199), Thomas (4 articles)

B. Examples of the position
Allen, Burlingame (30), Howland, Mesthene (162, 163), Price (194), Watson-Watt (7 articles)

4. Success-story format (pp. 145–146)

A. Critiques
Ferguson (70, 71), L. Hunter, H. M. Jones, Post (191), Rürup (6 articles)

B. Failure studies
Brown, Hoberman, L. Hunter, Jensen and Rosegger, Perkins, Post (191), Tascher (7 articles)

5. Western dominance (pp. 146–148)

A. Critiques
Buchanan, DeWalt, Ellul, Hartner, Rürup, Woodruff and Woodruff (6 articles)

B. Examples of the position
Brown, Burlingame (30), Heilbroner, Howland, Mesthene (162, 163), Watson-Watt (7 articles)

20. Technological momentum

1. Explicit use of "momentum" or "inertia" (pp. 149–154)
Bailes, Ferguson (70), Hounshell, Hughes (111), Kevles (132), Rosenbloom, T. Smith (7 articles)

2. **Unforeseen consequences of technology (p. 148 n. 55)**
   Ellul, Layton (141), Mesthene (162, 164), Pursell (199), Rosenberg (212), Rosenbloom, Rürup, Susskind and Inouye (9 articles)
3. **Enduring nature of existing technical concepts (pp. 155–156)**
   Bryant (22, 23), Cardwell (36, 37), Constant, Cowan (48), Fullmer, Harrison, Hilliard, Hoberman, Hounshell, Howard, Hughes (112), Jenkens, Kerker, Layton (144), Leslie, Mayr (156), Perkins, Reynolds, Sandler, Sinclair (226), Usher, Vincenti, Welsh (258), Wik (264) (26 articles)
4. **Enduring nature of existing technological artifacts (pp. 156–157)**
   Cain (33), Feller, Fries, Gade, Howard, Hughes (113), L. Hunter, Jensen and Rosegger, McCutcheon, Schallenberg and Ault, Sinclair (225), Tucker, Wilkinson (266), Znachko-Iavorskii (14 articles)
5. **Enduring nature of governmental policy (p. 157)**
   Bailes, Burke (27), Carver, Dorn, Hagen, Hughes (111), D. Jones, Perkins, Perry, Pursell (199), T. Smith, Tucker (12 articles)
6. **Enduring nature of corporate vested interests (pp. 157–158)**
   Bilstein, Burke (27, 28), Chapin, Clark, Dorn, Feller, Flick, Hoberman, Hounshell, Jeremy, Krammer, Leslie, Nicholas, Pursell (199), Rosenberg (211), Scherer, Sinclair (226), Strassmann, Wise (20 articles)
7. **Enduring nature of technological enthusiasm (p. 158)**
   Clarke, Feibleman, Ferguson (70), Hagen, Jensen and Rosegger, D. Jones, Kevles (132), Leslie, Perkins, T. Smith (10 articles)
8. **Enduring nature of cultural values (pp. 158–161)**
   Allen, Cowan (48), Dalrymple (49), Daniels, DeWalt, Drucker (59), Durbin, Esper, Feibleman (67), Fries, Gade, Hacker (91), Hagen, D. Jones, H. Jones, LaForce, Leighton (146), Lieberstein, Loria, Morris, Mumford (178), Pearson, Perry, Post (191), Pursell (199), Rosenbloom, Sandler, Shriver, Sinclair (226), Skolimowski, Susskind and Inouye, Tascher, Welsh (258), Uselding (34 articles)

Whig History and Technology and Culture *Authors (Chapter 5)*

21. Questions seldom discussed

1. **Failure studies (pp. 175–176)**
   A. Articles calling for failure studies
   Ferguson (70), L. Hunter, H. Jones, Post (191), Rürup (5 articles)
   B. Failures within longer successful traditions
   Bailes, Burke (27), Dalrymple (50), Feller, Frazier, Fries, Hughes (111), L. Hunter, Pearson, Post (191), Tascher, Woodbury (12 articles)
   C. Disappearance of or noninvestment in a technology
   Esper, Flick, Kevles (132), Leslie, Perkins, Tucker (6 articles)
   D. Miscellaneous perspectives
   Cowan (48), Jensen and Rosegger, Thomas (3 articles)
2. **Worker perspective (pp. 176–177)**

A. References to worker tensions
Brittain and McMath, Cowan (47), Dalrymple (49), Gorman, D. Jones, Mayr (155), Pursell (196, 199), Rasmussen, Susskind and Inouye, Thomas (11 articles)

B. Omission of worker perspective
Clark, Feibleman (67), Harris and Pris, Jeremy, Welsh (259), Wilkinson (266) (6 articles

3. Culture conflict in technology transfer (pp. 177–178)

A. References to colonialism
Brown, Gade, Gorman, Hoberman, Jensen and Rosegger, Leighton (146), McCutcheon (7 articles)

4. Studies of non-Western technologies (pp. 178–179)

A. Individual studies
Bray, DeWalt, Frankel, Sleeswyk, Sun, Wachsmann and Kay (6 articles)

B. Multiperiod surveys
Calder, Dresbeck, Fox, Kreutz, C. Smith (232, 233), White (261), Znachko-Iavorskii (8 articles)

5. Critiques of capitalism (pp. 179–180)

A. References to capitalist assumption in historical articles
Burke (27, 28), Dorn, Mayr (155), Thomas (5 articles)

B. Nonhistorical essays discussing premises of capitalism
Buchanan, Clarke, Durbin, Ellul, Hartner, Heilbroner, Mumford (176, 178), Rosenbloom, Shriver (10 articles)

6. Women and technology (p. 180)

Cowan (47, 48), Vanek (3 articles)

Notes

Chapter 1

1. Melvin Kranzberg to John Staudenmaier, 4 March 1983.

2. The brevity of my coverage of the "contextual" history of SHOT and TC is not due to a lack of appreciation for the sociological study of an emerging disipline. For the reasons mentioned in the introduction I am primarily interested in the new historical language that a careful text analysis finds in TC's articles. The reader may wish to consult my more detailed treatment of SHOT's institutional history in the dissertation from which this work has grown, "Design and Ambience: Historians and Technology: 1958–1977" (1980).

Some of the original dissertation and a number of its appendixes are not reproduced here. For reference the following list may be helpful. Appendix 1 focuses on TC's author constituency, including (1) a table listing by year all contributions to TC by author from 1959 to 1977; (2) a table listing each author's professional self-identification and country of residence; and (3) author responses to selected questions (personal interviews and questionnaire) about TC and the history of technology generally. Appendix 2 treats SHOT's relationships with other academic and technical societies: (1) joint sessions, at SHOT annual meetings, with other academic and technical societies through 1976; (2) SHOT joint sessions with other societies apart from annual meetings through 1970. Appendix 3 covers SHOT's membership growth from 1959 through 1976, and appendix 4 the society's outside sources of funding through 1976. Appendix 5 is a list of SHOT award winners—the Dexter, Usher, and Leonardo da Vinci prizes—through 1977.

All further citations of books and articles will be by short title, and the society and the journal will be referred to by their customary acronyms, SHOT and TC. Full citations may be found in the bibliography.

3. Most if not all of the early financial support was raised by Kranzberg. His relationship with the Kaufmann Foundation, by far SHOT's major source, is illustrative.

"I went to New York City and spoke with Edgar Kaufmann, Jr. (whose membership in SHOT dated back to 1960), and in 1964 the Foundation gave us $2,500. Although that enabled us to take care of our 1963 deficit, the deficits continued. In 1965, I approached Ed Kaufmann again because we had a $3,194.10 deficit for publication of our 1964 volume—and we anticipated further deficits for the next two or three years, since we had shifted to the University of Chicago Press from the Wayne State University Press. Taking a hint from Kaufmann, I applied for a three-year grant of

$5,000 from the Foundation for each of three years, beginning in 1965. So we received $15,000 as a result. . . . In 1969, they gave us $3,700, and in 1970 we received $4,000. We tried again in 1972 but they turned us down. However, you can see that the Edgar J. Kaufmann Charitable Foundation really helped us out in our struggling years by paying off our deficits—and enabling us eventually to accumulate a surplus." Kranzberg to Staudenmaier, 1 April 1983.

Case Western Reserve, where Kranzberg taught at the time, was another major contributor through the generous support services they extended to Kranzberg in his role as secretary of the society and editor of the journal. Leighton A. Wilkie, head of the DoAll Machine Tool Company of Des Plaines, Ill., also provided key early support through his Wilkie Brothers Foundation. For other funding sources see Staudenmaier, *Design and Ambience*, appendix 4.

4. The methods used for ascertaining who belonged to the leadership constituency of SHOT are explained in detail in Staudenmaier, *Design and Ambience*, appendix 1. Leaders who published at least one article in TC before 1963 were Silvio A. Bedini, L. Sprague de Camp, Carl W. Condit, Peter F. Drucker, A. Rupert Hall, Thomas P. Hughes, Melvin Kranzberg, Lewis Mumford, John B. Rae, Ladislao Reti, Cyril Stanley Smith, Lynn White, Jr., Reynold M. Wik, and Robert S. Woodbury.

For a list of all authors and their contributions to TC through 1977, see *ibid.*, appendix 1.

5. For a list of more recent active SHOT figures see ibid., p. 46 n. 45. Examples of key SHOT members who have not published an article in the journal include John Brainerd, Bern Dibner, Jack Goodwin, Joseph Jackson, Brooke Hindle, and Neil Fitzsimmons.

6. Evidence for the emergence of a modest backlog of manuscripts in 1963 can be found in the massive body of correspondence (hereafter "Correspondence"), which was on file in TC's editorial offices at the Georgia Institute of Technology, Atlanta, when I studied it. It was made available to me by the gracious assistance of Kranzberg and the editorial staff in April 1977. On the 1963 backlog, see Kranzberg to White, 15 October 1963.

7. Seventeen authors, responsible for twenty-two of the early articles, fall into this group. Nine identified themselves as historians of technology (R. S. Woodbury, L. White, Jr., T. P. Hughes, C. W. Pursell, R. C. Hall, J. B. Rae, R. Burlingame, L. Mumford, and M. Kranzberg), four as historians of science (W. Miles, F. Kilgour, S. Bedini, and W. Hartner), and four as historians of science and technology (C. W. Condit, A. R. Hall, R. P. Multhauf, and A. Zvorikine). For a full presentation of the professional self-definition of all authors, together with an explanation of the evidence used in making these judgments, see Staudenmaier, *Design and Ambience*, appendix 1.

8. The "founding meeting" was held in January 1958 at Case Western Reserve in Cleveland. Following the 1957 Cornell meeting with Condit, Rae, and Hughes, Kranzberg began to invite interested scholars to join a committee "for the promotion of the history of technology." The group included, among others, Lewis Mumford, Lynn White, Jr., William Fielding Ogburn, and David Steinman. Kranzberg describes the January 1958 meeting: "Most of the people on my committee could not come to the meeting. They were famous people, they were busy people, and they did not really have time to devote to this amorphous new effort. But some important people did come: Bob Multhauf, who at that time was head of the Science-Technology Division at the National Museum of the History of Technology [Smithsonian Insti-

tution]; Hugo Maier of Penn State, who did pioneer work in studying the relations between technological developments and American society; Carl Condit and John Rae, of course—and, most importantly, Lynn White!" Kranzber to Staudenmaier, 4 March 1983.

On SHOT's relations with other academic and technical societies, see Staudenmaier, *Design and Ambience*, appendix 2, and chapter 5, pp. 181–182.

9. The pattern held through 1980. Of the 224 authors published in TC during the period, 80 percent resided in the United States, 9 percent in Britain, and 3 percent in Canada. See ibid., appendix 1.

10. "I went after big names, men whose reputation alone was sufficient guarantee of quality. . . . I figured that if people with "big names" published in T&C, this would encourage others to enter the field. It lent a tone of intellectual respectability when well-known authors, who already commanded a large audience for their writings, gave heed to my importunings and sent me an article, or when somebody of David Riesman's stature agreed to serve as an Advisory Editor." Kranzberg to Staudenmaier, 1 April 1983.

The proceedings of The Encyclopaedia Britannica Conference on "The Technological Order" were published in full (TC 3, Fall 1962). See Kranzberg to Multhauf, 8 August 1962; Kranzberg to White, 6 February 1962 ("Correspondence"). This inclusive set was exceptional. The other single-theme issues—"Review Issue: A History of Technology" (TC 1, Fall 1960), "Science and Engineering" (TC 2, Fall 1961), and "The History of Rocket Technology" (TC 4, Fall 1963)—comprised contributions from various soures.

Journal correspondence indicates that at least the following SHOT members were active in soliciting manuscripts for TC: T. P. Hughes, R. P. Multhauf, C. S. Smith, L. White, Jr., R. Wik, and, of course, Kranzberg himself.

While some of these earliest articles were accepted without refereeing, TC correspondence indicates that refereeing began very early and that it became the universal practice after the first few years.

11. The correspondence of Lynn White, Jr., Cyril Smith, and John Rae was particularly extensive throughout the early period. All three discuss a wide range of societal and journal affairs with Kranzberg and others.

12. A summary of author responses is included in Staudenmaier, *Design and Ambience*, appendix 1. Because all respondents were guaranteed anonymity, it is impossible to cite any by name.

13. The founders recognized that the title of the journal would itself be a statement of intent, and the selection was not made easily. It was argued, for example, that "Technology and Culture" was so inclusive that it would frustrate any attempt to identify a specific disciplinary focus. Kranzberg, interview held at Georgia Institute of Technology, Atlanta, March 1977; R. P. Multhauf, interview held at the Smithsonian Institution, Washington, D.C., October 1977. See also Kranzberg, "At the Start," and the opening pages of chapter 4.

14. The elevan historiographical articles of this period reveal a striking bias in favor of broad contextual studies. No article argues for a pure internalist approach, and a number explicitly criticize it. See especially Allen, "Social Change"; H. Jones, "Ideas"; Kranzberg, "At the Start"; Mumford, "Neglected Clue"; White, "Invention." Two articles, unique in TC, provide a glimpse of the Marxist critique of pure internalism. They are balanced by a third questioning the validity of Marxist practice

in the Soviet Union. The first two are by Zvorikine, "Soviet History" and "Technology." The critique is Joravsky, "Marxism."

15. Kranzberg, "At the Start," p. 7.

16. Mumford, "Neglected Clue," pp. 230–231.

17. The three, together with a commentary by Derek de Solla Price, originated as a rare historiographical session at SHOT's 1972 annual meeting.

18. The other historiographical articles published in the seventies are: Daniels, "Big Questions" (comments by J. G. Burke and E. Layton); Mayr, "Historiographic Problem"; White, "Reflections"; Keller, "Tortelli"; and Hughes, "Themes." Daniels treats many of the same themes that are discussed here under the rubric of the 1974 collection of articles. White and Hughes stress many of the same issues as Rürup. Keller and Mayr treat more specialized questions.

The 1978 special SHOT conference, "Critical Issues in the History of Technology," held in Roanoke, Va., in August 1978, and the Hughes 1979 article appear to be the beginning of renewed historiographical interest in recent years. David Hounshell ("History of American Technology," 1980–1981, and "Letter," 1981–1982) and Darwin Stapleton ("Letter," 1981–1982) exchanged views on the state of the field in the pages of the *Journal of American History*. John Staudenmaier, "What SHOT Hath Wrought" (1984), together with responses by Kranzberg and Rae, has already sparked further reflections. See Brooke Hindle, "Exhilaration" (1984).

19. Ferguson, "Toward a Discipline," pp. 13–18; Multhauf, "History of Technology," pp. 7–8.

20. Multhauf, "History of Technology," p. 9.

21. Ferguson, "Toward a Discipline," pp. 21–30; Rürup, "History of Technology," pp. 183–191. See also Daniels, "Big Questions," pp. 2–21; Hughes, "Themes," passim.

22. Rürup, "History of Technology," p. 193.

23. Leading members of the society continued to debate the quality of historical research and technical competence in TC's articles. This appears to have been one of several related questions discussed in the stormy executive council meetings of SHOT's annual meeting in December 1973. See TC 15 (July 1974): 472.

24. Kranzberg argued that the authors, by their choices of method and research topics, must define the identity of the field. For a further discussion of TC's policy of publishing a broad range of scholarship see Staudenmaier, *Design and Ambience*, pp. 24–26.

25. For a more detailed analysis of the difficulty of this integration see Colleen A. Dunlavy's "Transcending Internalism: On the Historiography of Technology in the United States" (unpublished manuscript written for the Science, Technology, and Society Program at MIT, February 1983). Dunlavy advances powerful evidence that SHOT members have remained locked in the internalist approach at least through 1981 and that genuine integration of design and context is exceedingly rare in SHOT circles. See note 63 below.

26. On Tortelli, see Keller, "Tortelli," pp. 345–365. On von Poppe, see Multhauf, "History of Technology," p. 74. Rürup's survey of the "prehistory" of the history of technology ("History of Technology," pp. 167–168) ignores the early histories of inventions and, curiously, von Poppe. He begins the prehistory with the eighteenth-century French *Encyclopédie* and the twenty-volume *Descriptions des Arts et Métiers* on

the one hand and Johann Beckmann's approach to the "science" of *Technologie* (Göttingen, ca. 1770) on the other.

In this context we should note the parallel European tradition of science museums. The great museums that flourished in seventeenth-century Germany, Italy, and England served as research facilities, as scientific libraries, and as incipient artifact encyclopedias. They were themselves an outgrowth of a much earlier tradition of collections of curiosities such as that of Jean de Berry, Duke of Burgundy (1340–1416). For a survey of science museums, see Bedini, "Museums."

27. *Beiträge* ceased publication in 1941 and reemerged as *Technikgeschichte* in 1965 (Düsseldorf). *Archiv* ceased publication in 1931. While *Beiträge* devoted itself to the history of industrial technology, *Archiv* increasingly concentrated on the history of science. See Rürup, "History of Technology," pp. 168–169. Rürup remains the most helpful survey of European scholarship. For other discussions of extant literature in the field, see Multhauf, "History of Technology"; Ferguson, "Toward a Discipline"; Derek de Solla Price on Multhauf and Ferguson, "Commentary"; and the entire fourth number of TC 1 (Fall 1960), which is dedicated to reviews of recently published histories of technology.

28. For a survey of research into the impact of technology on society in the United States, see Allen, "Social Change."

29. Charles Singer, E. J. Holmyard, and A. R. Hall, eds., *History of Technology* (1954–1958); Maurice Daumas, ed., *Histoire* (1962–1968); A. Zvorikine, *Geschichte* (1962).

30. See A. Hall's reflections introducing TC's collection of review essays for the Singer five volumes (Fall 1960), "History of Technology," p. 314.

31. Singer et al., *History of Technology*, 1: vii.

32. In particular see Multhauf, "History of Technology," pp. 4ff., and A. R. Hall, "History of Technology," p. 314.

33. Schmookler, *Invention* (1966). For articles prefiguring this major work see "Changing Efficiency" (1952); "Patent Statistics" (1953), "Inventive Activity" (1954); Ogburn, *Social Change* (1923); Gilfillan, *Sociology of Invention* (1935).

34. They disagreed sharply in their interpretation of the role of patent statistics for determining inventive activity. See Gilfillan, Schmookler, and Kunkik, "Patents."

35. Hunter, *Steamboats* (1949); White, *Medieval Technology* (1962). Other noteworthy examples of early contextual books include Taylor, *Transportation Revolution* (1951); Holley, *Ideas and Weapons* (1953); Josephson, *Edison* (1957); and Aitken, *Taylorism* (1960).

36. Mumford, *Technics* (1934). On Mumford's influence among early SHOT leaders see the beginning of chapter 5.

37. Usher, *Mechanical Inventions* (1929).

38. For a complete discussion of the taxonomies used here see appendix 2.

39. Seven of the nine historiographical articles after 1970 were written by leading SHOT scholars (i.e., Ferguson, Hughes, Layton, Mayr, Keller, Multhauf, and White). See note 18 above.

40. Gilfillan et al., "Patents."

41. Two-thirds of the nonhistorical essays either omit source notations altogether or refer only to secondary sources, another indication of their lack of engagement in critical discourse with TC readers.

42. It should be noted that my scoring of contextual articles is more broadly based than some critics prefer (see Dunlavy, "Transcending Internalism," pp. 15-16). I have scored as "contextual" all articles that situate a technological design in some larger context, whether that context includes major "patterns of human social interaction" (Dunlavy, p. 50) or only more individualistic factors and indeed economic factors that implicitly assume the social validity of the "free market" economy. This broad definition of the contextual style has been chosen to isolate pure internalist articles and to make visible the gradual influence of the contextualist ideal in the field. The difficulty of a genuine contextual style, which demands an integration of technical design and social context and not mere juxtaposition, should not be underestimated.

For my own critique of the field's avoidance of critical social analysis see the opening pages of chapter 4, where the inchoate and underdeveloped status of the relationship between technical design and culture is more thoroughly discussed, and especially chapter 5, passim, for a full-scale analysis of TC's success in achieving genuine contextualism.

43. Dornberger, "German V-2."

44. Woodbury, "Whitney Myth," p. 381, and White, "Malmesbury," p. 97.

45. White, "Invention," p. 486.

46. Boyer, "Axles." See appendix 2 for other internalist articles adopting the *a priori* style.

47. Interaction among internalists took place not only in TC's articles but even more frequently in published correspondence and other items under the rubrics of "Documents," "Research Notes," and "Controversy." Typical is A. Rupert Hall's comment ("Axles") on Boyer's discussion of medieval pivoted axles, "Axles." He supports her conclusions with his own evidence found in medieval author Guido da Vigevano.

48. For discussion of evidence that internalist historians have been influenced by TC's contextual style see Staudenmaier, *Design and Ambience*, pp. 82–85.

49. Internalist articles hold at 18 percent in the middle years and dip to 16 percent in the most recent period. An untypical decline in *a priori* articles in the middle period (to 46 percent) is followed by a return to dominance in the most recent seven years (79 percent). See appendix 2.

50. Brown, "Ewo Filature"; DeWalt, "Appropriate Technology"; Feller, "Cotton Industry"; Fries, "British Response"; "Hacker, "China and Japan"; Hoberman, "Desague"; Jeremy, "Textiles"; Jensen and Rosegger, "Lower Danube"; Roland, "Bushnell"; and Ruttan and Hayami, "Transfer."

51. Bryant, "Diesel"; Constant, "Turbojet"; Hanieski, "Airplane"; Harrison, "Single-Control Tuning"; Hewlett, "Nuclear Development"; Hounshell, "Telephone"; Hughes, "Electrification"; T. Smith, "Whirlwind"; Uselding, "Forging"; Vincenti, "Tests"; and Wise, "GE Research."

52. Hughes, "Power Transmission"; Jevons, "Mother of Invention"; Layton, "Ideologies" and "Turbine"; Mayr, "Porter Engine"; Multhauf, "Salt"; Reynolds, "Waterwheel."

53. Bailes, "Soviet Aviation" (technology as a legitimating factor); Burns, "Greek Water Supply" (ancient Greek urban engineering); Cowan, "Women and Technology" (American women and technology); Emmerson, "Great Eastern" (critique of Russell's

version of the affair); Ferguson, "American-ness" (survey of American technological style); Garcia-Diego, "Weirs" (medieval dam construction); Howard, "Interchangeable Parts" (interchangeable parts hypothesis reconsidered); Rosenberg, "Interdependence" (American infrastructural relationships); Schallenberg and Ault, "Charcoal Iron" (earlier explanation of decline of charcoal iron industry critiqued) TeBrake, "Pollution" (concern over pollution).

54. Mayr, "Historiographic Problem," p. 664.

55. See Staudenmaier, *Design and Ambience*, p. 71.

56. Kevles, "Postwar Research."

57. Cognitive anthropology was introduced to me at the University of Pennsylvania where leading scholars such as Anthony Wallace and Ward Goodenough have fostered the approach. For an excellent collection of cognitive articles, see Spradley, ed., *Culture and Cognition* (1972). For the general epistemological philosophy I am indebted to George Klubertanz, S.J., and James Collins, at the St. Louis University Philosophy Department where I completed a master's degree in 1964. For the hermeneutical approach I follow here, see Gadamer, *Truth and Method* (1975).

58. The example is based on a contrast between two articles, Robinson, "Watt," and Kerker, "Steam Engine."

The convention s/he will be used whenever the third person singular pronoun is required.

59. For a thorough analysis of the demands of verification, see Murphey, *Historical Past* (1973).

60. Spradley, ed. *Culture and Cognition*, p. 191. For a sociological parallel to the anthropological theory that follows, see Berger and Luckmann, *Social Construction* (1966).

61. The principle has been articulated in Hage, "Beer Categories," p. 269, in Spradley, ed., *Culture and Cognition.*

62. In cognitive anthropology it is possible to distinguish two levels of an outsider's understanding of a culture, namely: "structually valid" understanding (where the outsider's understanding adequately accounts for the normal rules of behavior but without an understanding of the cognitive universe that grounds that behavior) and "psychologically valid" understanding (where the outsider has come to understand the cognitive world view of the culture as well as the behavior it governs). In the text here I am referring to the second level of "psychological validity" when I speak of achieving fluency in the culture's language. Of course, any child born into a culture must achieve this same fluency. The distinction is taken from course notes of Dr. John Caughey of the University of Pennsylvania, 1973.

For a discussion of achieving fluency in science similar to the idea presented here, see Price, "Technology Independent," p. 558.

63. Colleen A. Dunlavy has cogently argued that this process—of the culture's historical universe of discourse influencing the assumptions of historians—operated among SHOT historians to foster the internalist bias.

"The established historian [at SHOT's beginning] came from somewhere—in fact, from two places: from one of the older branches of history and from the world at large. In both places, 'the pre-existing set of assumptions about the meaning and shape' of history was presumably more settled and coherent. In the world of American history in the 1950's, the tenor was generally set by the 'consensus school' of

American historians, while in the [American] world at large 'ideology' seemed to have died after the turmoil of the 1930's, World War II, and McCarthyism—which is only to say that a conservative tenor prevailed. It seems reasonable to think that both phenomena contributed in the late 1950's and early 1960's to the 'cultural ambience' of the historian, whether in an established field or recently arrived in a new one. If one notes that this conservative atmosphere inside and outside the historical world was remarkably congruent with the internalist's theoretical approach to history, then the 'cultural ambience' of the times presumably lent its force in subtle ways to the advocates of the internalist perspective in the history of technology." "Transcending Internalism," pp. 14–15.

64. Gadamer, *Truth and Method*, p. 324.

65. H. Jones, "Ideas," p. 22. Brooke Hindle makes the same point in *Emulation* (1981).

66. See Dunlavy, note 63 above.

67. One other significant measure of evidentiary priorities, the use of visual materials in TC articles, has been treated in appendix 2 of Staudenmaier, *Design and Ambience*. Given the influence of publication cost constraints on such materials, it is difficult to draw conclusions about author methodologies from these patterns, and they have been omitted here.

68. On consensus about development see chapter 2, pp. 45ff. On contradictory interpretations of transfer see Chapter 4, pp. 128–134. Throughout chapters 2, 3, and 4 I will try to indicate what degree of consensus exists in author usage about each thematic interpretation.

69. If we omit the references that are not of central interest in the article, the pattern is even more striking. The United States and Europe account for almost 90 percent of such references and the Middle Eastern and Mediterranean basin references increase the Western bias to an overwhelming 95 percent. Thirty-eight articles make no reference to a specific place. Not surprisingly, they tend to be nonhistorical and historiographical essays.

70. For more details about articles published before 1977 relating to technological topics as well as time and place boundaries, see Staudenmaier, *Design and Ambience*, pp. 99–120 and appendixes 9, 10, and 11.

71. It is interesting to note that this composite pattern (from the Middle East to contemporary United States) is mirrored almost exactly in several articles that trace broad themes through long time periods, for example; Burlingame, "Neglected Clue," and Finch, "Engineering and Science."

72. Burns, "Greek Water Supply"; Hacker, "Greek Catapults"; Kahn, "Greek Tragedies"; and Poole and Reed, "Dead Sea."

73. The bulk of the nonhistorical articles fall, as would be expected, in the set of articles with no date references. The twentieth-century nonhistorical articles have been dated in the century because their authors focus attention on technological issues peculiar to the recent past. The historiographical articles show a similar pattern.

Chapter 2

1. White, "Invention," p. 498. The term "innovation" is sometimes found in the literature referring to the third stage of the invention–development–innovation process

and sometimes as a cover term for the entire process. To avoid this ambiguity I have chosen the expression "emerging technology" as a cover term for the whole process and will restrict "innovation" to its more limited use.

2. Charles O. Frake, among others, indicates that the role of linguistic usage in determining the structure of a culture's cognitive universe is the central tenet of cognitive anthropological theory: "Linguistic forms, whether morphemes or larger constructions, are not each tied to unique chunks of semantic reference like baggage tags; rather it is the *use of speech*, the *selection of one statement over another* in a particular socio-linguistic context, that points to the *category boundaries* on a culture's cognitive map" ("Cognitive Systems," in Spradley, ed., *Culture and Cognition*, p. 195; my italics).

3. The three articles adopting a Marxist perspective are Zvorikine, "Soviet History" and "Technology," and Joravsky, "Marxism." All three are historiographic essays. No historical study in TC adopts a Marxist perspective during the years covered by this study.

4. White, "Invention," p. 497.

5. It should be noted that the frequencies presented in the table are not considered rigorously valid in the statistical sense. The scoring of the articles on thematic taxonomies is necessarily based on my interpretative judgments and is therefore subject to further correction. Thus the frequencies should be regarded as an approximation of TC usage. See appendix 1, where the methodology followed in designing and scoring thematic taxonomies is discussed.

6. Reti, "Martini," p. 292.

7. Bryant, "Four-Stroke Cycle," p. 186.

8. Brittain, "Loading Coil," pp. 42–44, and Robinson, "Watt," p. 128.

9. White, "Invention," p. 487.

10. Agassi, "Confusion." In this context it is interesting to note psychologist Rollo May's suggestion that the avoidance of reflection on the nature of the creative process is not limited to historians. He critiques Alfred Adler's "compensatory theory of creativity" ("that human beings produce art, science, and other aspects of culture to compensate for their own inadequacies") in the following terms. "Its error is that it does not deal with *the creative process as such.* Compensatory trends in an individual will influence the forms his or her creating will take, but they do not explain the process of creativity itself" (*Courage*, pp. 37–38).

11. Mayr, "Porter Engine," p. 570.

12. Hounshell, "Telephone"; C. S. Smith, "Art"; Watson-Watt, "Modern World"; and White, "Malmesbury," p. 101.

13. Fox, "Fire Piston," pp. 359–360.

14. Jenkins, "Eastman." The concept of intellectual background as the substratum for an inventor's interaction with the state of the art is closely related to the technical tradition (see p. 65) and the systems approach (see pp. 69–72).

15. Chapin, "Patent Interferences," p. 443.

16. Finch, "Engineering and Science," p. 328.

17. Bryant, "Diesel," p. 444.

18. Dornberger, "German V-2," p. 399.

19. Scherer, "Watt-Boulton," p. 175.

20. Bryant, "Diesel," p. 444.

21. For example, Post, "Page Locomotive."

22. Leslie, "Copper-Cooled Engine," and Perkins, "DDT." Fourteen articles focus on a military context and fifteen on corporations. Both types of institution have drawn increasing attention from historians of technology. See, for example, Chandler, *Visible Hand*, and Noble, *America by Design*, for corporate influences. On the role of the military see Merritt Roe Smith, ed., *Military Enterprise*, especially his introduction and Alex Roland's bibliographical essay.

23. This is, of course, an oversimplification of feedback mechanisms, which are ordinarily designed to respond to the momentum that carries the system past the desired level by anticipating the appproach to that level and setting corrective measures in motion before the system reaches it. For our purposes, however, the more important aspect of a feedback loop lies in the error-correction signal amid the physical momentum of the system rather than in those adjustments aimed at diminishing the effects of momentum.

24. Since no device is an absolutely perfect embodiment of its design concept this feedback correction process is endless in principle. In practice, however, the constraints of time, costs, and market strategy will eventually limit it.

25. Perry, "Atlas," pp. 470–471.

26. R. C. Hall, "Satellite," pp. 416–417.

27. T. Smith, "Whirlwind," pp. 456–458.

28. Hughes, "Power Transmission." On the definition of the problem see pp. 647–648; on subsequent experimentation see pp. 650ff; on the role of technical journals see p. 649.

29. Bryant, "Silent Otto," p. 198.

30. Two recent articles address the problem directly: Vincenti, "Tests," and Wise, "GE Research." Vincenti calls attention to the distinction between development of "the technological base" as opposed to "development of a specific project" (p. 750). Wise notes that "distinctions among research, development, and engineering were then, [ca. 1900 at General Electric] and remain ambiguous" (p. 427).

31. Bryant, "Diesel," p. 446.

32. Constant, "Turbojet," p. 554.

33. Ibid., p. 557.

34. Ibid., p. 558. Constant's reflection on the aesthetic appeal of the paradigm also occurs on p. 558.

35. Post, "Page Locomotive."

36. Scherer, "Watt-Boulton." Schumpeter's five characteristics of the innovator, as summarized by Scherer, are: "1. New combinations are as a rule embodied in new firms which do not arise out of the old ones but start producing beside them. 2. The entrepreneur–innovator is characterized by 'initiative,' 'authority,' and 'foresight'; he is the 'captain of industry' type. 3. The only man the entrepreneur–innovator has to convince is the banker who is to finance him. 4. The entrepreneur–innovator retires from the arena only when his strength is spent. 5. The entrepreneur–innovator's motivation includes such aspects as the dream to found a private kingdom, the will to conquer and to succeed for the sake of success itself, and the joy of creating and getting things done" (p. 172).

37. Bryant, "Diesel," p. 445.

38. Jeremy, "Textiles." For a contrast to Jeremy's purely economic interpretation of Lowell, see Kasson, *Civilizing the Machine*, chap. 2.

39. But see Hounshell's study of Bell's breakthrough invention for an interpretation suggesting a significant role for economic motives at the inventive stage ("Teleophone," p. 156).

40. Jenkins, "Eastman," esp. p. 13.

41. Chapin ("Patent Interferences") is most thorough.

42. D. Jones, "Tear Gas," p. 159 and passim.

43. The term "diffusion" suffers in TC from the ambiguity of dual usage: diffusion within a single culture and diffusion across cultural lines. The second meaning is more commonly called "transfer of technology." For a distinction between the two see chapter 4, p. 123.

44. Rosenberg, "Transfer," p. 566. Note that studies by Sinclair ("Screw Thread") and Clark ("Car Coupler") are excellent examples of Rosenberg's point.

45. M. R. Smith, "Hall and North."

46. Ruttan and Hayami, "Transfer," p. 132; Hughes, "British Lag"; Feller, "Cotton Industry."

47. Examples of articles that discuss all three concepts are Bryant, "Diesel"; Chapin, "Patent Interferences"; Constant, "Turbojet"; Mayr, "Porter Engine"; and Scherer, "Watt-Boulton."

48. Keller, "Tortelli," pp. 346–347.

49. On the problem of plagiarism, see Reti, "Martini," and Bedini, "Clepsydra." Robinson ("Watt") and Post ("Patent Office") study aspects of the development of patent law after 1800. Multhauf ("Improver") notes the impact of patent law on the definition of the term "inventor."

50. Usher, *Mechanical Inventions*, Gilfillan, *Sociology of Invention*, and Ogburn, *Social Change*. On Usher's relationship with the internalist tradition, see p. 12. The companion volume to Gilfillan's *Sociology of Invention* is titled *Inventing the Ship*. For Ogburn's dependence on the internalist tradition, see *Social Change*, pp. 73–102 and 200–212. For internalist articles verifying inventive claims, see appendix 3.1.1.

51. Schumpeter, *Economic Development*, pp. 84–98. For a list of Schmookler's works, see chapter 1, note 33.

52. See especially Chapin, "Patent Interferences," pp. 438–446.

53. Scherer, "Watt-Boulton," pp. 186–187.

54. Rosenberg, "Transfer," p. 569.

55. For example, Hughes articulates the invention-development-innovation model in *Sperry*. His work was independent of Scherer's. Interview with Thomas P. Hughes, February 1980.

56. Bryant, "Silent Otto" and "Diesel." Dornberger, "German V-2," is another example of an early description of developmental activities without the use of the term: "Modern invention is hard, scientific, and technical work by a whole group of intelligent, dedicated people. It is a matter of progressing step by step, examining the testing in different institutes and facilities, weighing the feasible against the

hoped for in many proposals" (p. 399). See Rosenberg, "Transfer," p. 569, for another example.

57. Hughes, "Development," pp. 423–424.

58. Heilbroner, "Machines," p. 339.

59. Cardwell, "Joule," p. 685.

60. Dornberger, "German V-2," p. 401.

61. Usher, "Industrialization," pp. 114–115.

62. Sinclair, "Screw Thread," p. 30.

63. Dalrymple, "Storage of Fruit," p. 35.

64. Cain, "Chicago Sanitation," p. 361.

65. Constant, "Turbojet," p. 554. For other explicit definitions see appendix 3.5.1.

66. Frazier, "Current Meter," p. 563.

67. Kerker, "Steam Engine," p. 383. The quotation is taken from Usher, *Mechanical Inventions*, p. 347.

68. Multhauf, "Saltpeter," p. 163.

69. Bryant, "Thermodynamics," pp. 155–156.

70. Dalrymple, "Storage of Fruit," p. 40. Mumford makes the same point in "Neglected Clue," p. 235.

71. Feibleman, "Skills," pp. 323–324.

72. See, for example, Hounshell, "Telephone."

73. Bedini, "Clepsydra."

74. Two other examples similar to Kerker, "Steam Engine," are Constant, "Turbojet," and R. C. Hall, "Satellite."

75. Bryant, "Silent Otto." For other examples, see the list of breakthrough invention studies—except Constant, Hall, and Kerker—in appendix 3.5.3.

76. Pearson, "Peacemaker," p. 182.

77. McLuhan and Nevitt ("Causality") comment on the tendency to interpret events as end points: "Donald A. Schon notes that 'tendency either to obscurantize or to explain away novelty reflects the great difficulty of explaining it. The difficulty comes in large part from our inclination, with things and thought alike, *to take an after-the-fact view*.' What the scientist normally sees is either a replay of past scientific experience or an up-and-coming threat in his 'rear-view mirror' " (p. 5; my italics). The citation from Schon comes from his *Displacement of Concepts*.

78. Adam Smith, *Principles*, p. 84, cited by Mayr, "Adam Smith," p. 17. Mayr devotes the entire article to an analysis of the interrelationship between Smith's theoretical model and contemporary technological systems.

79. Drucker, "Work and Tools," p. 36. For an unusually explicit and detailed discussion of the systems approach for interpreting technological change, see Rosenberg, "Interdependence," pp. 29–32.

80. Hughes, "Power Transmission," p. 646.

81. Constant, "Turbojet," p. 555.

82. The interpretation of systems thinking I have adopted here deliberately stresses the link between systems theory and the mechanistic bias toward standardization,

uniformity, and hierarchical control. My reading of TC usage suggests, as will be seen below, that the two tend to be linked in practice even though they do not necessarily have to be. Thomas Hughes's very recent attempt to move beyond a mechanistic approach (see his introduction to *Networks*, p. 6) is a significant step, by a SHOT leader, in this direction. On the same point see my chapter 5, p. 191. See also note 90 below.

The tension between hierarchical control and peer negotiation in Western cultures and especially in cultures permeated with industrial capitalism has become the subject of considerable interest in two scholarly areas. Labor historians such as E. P. Thompson (*English Working Class*), David Montgomery (*Workers' Control in America*), and Harry Braverman (*Labor and Monopoly Capital*) stress the transformation of the work place from what Shoshanah Zuboff calls "traditional society with all of its rich and boisterous rhythms of human interaction" to the "disciplines routine of factory work" ("*work and Human Interaction*," pp. 5–6).

The relatively recent emergence of gender studies in the United States provides a second perspective. Carol Gilligan speaks of the tension between the previously normative view, which defines moral adulthood in terms of hierarchy and autonomy, and the "different voice" of woman respondents, which addresses moral questions in terms of relationship and caring (*Different Voice*, e.g., p. 32). Again in Rothschild, ed., *Machina ex Dea*, we find the tension between masculine and feminine stereotypes commonly referred to with the hierarchy–negotiation dichotomy. See especially the essays by Merchant, King, Keller, Bush, and the more strident Gearhart.

The topic is too complex to be fully studied here, but see chapter 5's treatment of TC's avoidance of worker perspective, critiques of capitalism, and the women's perspective for some discussion of related materials.

83. Bryant, "Silent Otto."

84. Miles, "Polaris," p. 482.

85. Jeremy, "Textiles," p. 46.

86. Ibid., p. 48.

87. Both paragraphs, ibid., pp. 72–73.

88. For a perceptive discussion of the emergence of this systemic and standardized model of rail networks and its replacement of a previous negotiation model see Larson, "Systems Approach."

89. Clark, "Car Coupler."

90. The option for systemic standardization over network designs that depend on negotiation is, to be sure, a value-laden choice. Although it is often and rightly linked to the ideology of industrial capitalism, its military origins should not be overlooked. See M. R. Smith, "American System."

91. Carver, "Russian Journal," and Lieberstein, "Soviet NOT Movement."

92. Kohlmeyer and Herum, "Agriculture," p. 372.

93. Rosenberg, "Transfer," pp. 568–569.

94. I have found only four articles that refer to the impact of emerging technology on a system. Like the references to the impact of emerging technology on technical traditions, they are passing remarks with little thematic articulation. See appendix 3.6.8.

95. Theobald ("Emerging Nations") exemplifies this mentality when he cites the following from Senate Testimony of the Stanford Research Institute, "Possible Non-

military Scientific Developments and Their Potential Impact on Foreign Policy Problems of the United States," in *United States Foreign Policy* (Washington, D.C., 1961), p. 108: "Someone has said that in recent times *we have invented the art of systematic invention.* Organized scientific research and development, which has become *a great industry* in the last few decades, is itself one of the most significant social inventions of the twentieth century. It is unlocking the secrets of nature and putting the knowledge to practical use at an unprecedented rate. Also *we have invented the art of systematic innovation*" (p. 607; my italics).

96. Agassi, "Confusion," p. 353.

97. Chapter 5, pp. 173–181 in particular. See also chapter 3, pp. 95–103, and chapter 4, pp. 134–148.

Chapter 3

1. TC 17 (October 1976).

2. Arnold Thackray, comment in published discussion of Multhauf, "Salt," p. 645. Reingold and Molella, "Science and Technology," p. 632. Mayr, "Historiographic Problem," p. 668. Robinson, comment in published discussion of Hughes, "Power Transmission," p. 661. For other questions about the value of the science–technology model see Reingold, "Bache," p. 174, and Overfield, "Bessey," pp. 178–179.

3. Here and throughout the chapter my references to the methodological canons of the history of technology are intended to be a shorthand reference for the explanation of the contextual style of history presented in chapter 1. Of particular importance is the governing focus of the contextual approach which attempts to study the design characteristics of technology as they are related to specific historical contexts. It should be recalled that the argument for stating that the design-ambient tension establishes these governing methodological canons is not only based on the emergence of contextual articles as the majority of all articles after 1964, but also on the methodological dependence of externalist articles on the contextual style, and on the stated intent of SHOT to publish a journal studying technology *and* culture.

4. Many of TC's references to technological knowledge have been discussed as parts of chapter 2's theme, emerging technology. It will be recalled that cognition was a major element in many of the subthemes treated there. Technological knowledge is, however, more than a subordinate element of emerging technology. As we shall see, it is discussed not only with reference to new technologies but in other contexts as well.

5. Hewlett, "Nuclear Development," p. 470.

6. Bunge, "Applied Science," p. 329.

7. C. Smith, "Metallurgy," passim.

8. Hughes, "Power Transmission," p. 660. See also Reingold, "Bache," p. 174, and Meeker, "Art," passim.

9. Agassi, "Confusion," p. 358.

10. Ibid., p. 360. For another philosophical argument making the same point, see Skolimowski, "Thinking," esp. pp. 374, 377.

11. C. Smith, "Metallurgy," pp. 361–362. Cardwell makes the same point in "Power Technology," p. 205.

12. Layton, "Mirror-Image," p. 569.

13. Brittain, "Loading Coil," p. 39.

14. Bryant, "Thermodynamics," p. 157.

15. Mayr, "Porter Engine," pp. 591ff.

16. Multhauf, "Saltpeter," p. 176.

17. Ibid.

18. A. R. Hall, "Technical Act," p. 511.

19. Bedini, "Automata," p. 32.

20. Dryden, "Outer Space," p. 115.

21. Rae, "Know-How," pp. 143–144.

22. Bryant, "Silent Otto," pp. 195ff., "Diesel," p. 435.

23. Cardwell, "Power Technology," p. 188.

24. Buchanan, "Exploitation," p. 537.

25. Layton, "Mirror-Image," pp. 576–578.

26. Multhauf, "Salt," p. 642. See Overfield, "Bessey," pp. 178–179 for precisely the same point.

27. Reingold, "Bache," p. 175; Overfield, "Bessey," pp. 178–179.

28. Mayr, "Historiographic Problem," p. 670.

29. For more on the Hindsight–TRACES controversy, see Layton, "Mirror-Image," p. 564.

30. Bunge, "Applied Science," p. 330.

31. Ibid., p. 336.

32. Ibid., p. 338. For other authors who explicitly distinguish premodern craft from modern scientific technology see appendix 3.12.2.

33. Ibid., p. 339.

34. Ibid., p. 331. Bunge's position is remarkably close to the Marxist explanation of the science–technology relationship as articulated by Zvorikine in "Soviet History," p. 1.

35. Ibid., p. 332.

36. Morris, "Context," p. 412.

37. Ibid., pp. 410–411. On this point of science's triumph over earlier empiricism see also Bunge, "Applied Science," p. 330.

38. The authors who adopt some or all of the assumptions congruent with the applied science hypothesis have the following professional backgrounds. Mario Bunge, James K. Feibleman, and Bertram Morris are philosophers. Henry Leicester and Sir Robert Watson-Watt are scientists. John Lienhard is a mechanical engineer. A. Zvorikine is a Marxist historian of technology. Robert L. Heilbroner is an economic historian. A. Rupert Hall and Daniel J. Kevles are historians of science. Two historians who are closer to the mainstream of SHOT are unusual: John B. Rae and Reinhard Rürup use language suggesting the applied science model, but both also argue against

it in other sections of their work. See Rae, "Know-How," pp. 141, 143, and Rürup, "History of Technology." p. 186.

39. See, for example, Cardwell as cited above under statement 4. See also Reingold and Molella, "Science and Technology," p. 627.

40. C. Smith, "Art," p. 497. For a nonhistorical discussion see Meeker, "Art," passim. On Rudolph Diesel's aesthetic motives in design of technology see Thomas, "Diesel," p. 382.

41. C. Smith, "Art," p. 498.

42. Ibid., p. 533.

43. Ibid., p. 546.

44. Joseph Agassi ("Confusion," pp. 356–359) makes a similar argument when he notes that scientific theories are algorithms but that the verification or falsification of algorithmic systems does not lead to the creation of new or competing theories. In the act of scientific creation, other forces, such as Smith's aesthetic impulse, must account for the *emergence* of radically new theory. See also our discussion of Constant's theory of revolutionary technological paradigms in chapter 2.

45. Such critiques are commonplace in recent history of science where the claims of objective and value-free scientific practice are challenged by studies of the political and economic factors influencing the content and methodology of science. See Thackray, "History of Science," in Durbin, ed., *Science, Technology, Medicine*, and MacLeod, "Changing Perspectives," in Spiegel-Rösing and Price, eds., *Science, Technology, Society*. For studies of psychological and philosophical influences on the content and method of science, recent "gender and science" literature is particularly important. Keller has developed this perspective most thoroughly and insightfully (*McClintock* and *Gender and Science*). See also Rothschild, ed., *Machina ex Dea*, and Trescott, ed., *Dynamos and Virgins*.

46. Skolimowski, "Thinking," p. 373.

47. See note 3 above.

48. T. Smith, "Whirlwind." pp. 462–463.

49. Many TC historians would prefer here to use the term "engineering theory" to distinguish what Smith calls "indigenous science" from scientific theory. See the section "Engineering Theory" below.

50. Kerker, "Steam Engine," p. 383.

51. Skolimowski, "Thinking," p. 374.

52. Mayr, "Porter Engine," pp. 594–595.

53. Clark, "Car Coupler," pp. 182ff.

54. Post, "Page Locomotive," p. 164. The quotation within the text is taken from Hatfield, *Inventor*, pp. 72–73. Stuart W. Leslie's study of the General Motors attempt to develop an air-cooled engine in the early twenties is a remarkably thorough study of the influence of such motives as pride, jealousy, and vested interests within GM on interpretations of test data. See "Cooper-Cooled Engine," pp. 761–762, 764–765, 770, 773, and 775. See chapter 2, "The Process of Development," for other examples of biases in the structure of the data-gathering process.

55. Burke, "Bursting Boilers," passim, and Skramstad, "Canal Incline," p. 556.

56. Cain, "Watering a City."

57. Layton, "Ideologies," p. 695.

58. Constant, "Turbojet," pp. 555–556.

59. Hanieski, "Airplane," pp. 540–541.

60. Hughes, "Power Transmission," pp. 647–648.

61. Ibid., p. 649.

62. Ibid., pp. 653–654.

63. Ibid., pp. 646–647.

64. Brittain and McMath, "Georgia Tech," p. 177.

65. Rürup, "History of Technology," p. 186.

66. Feibleman, "Skills," p. 327.

67. Dornberger, "German V-2," p. 400.

68. Donald A. Schon presents a provocative analysis of the tension between experiential skill and theoretical deduction when he contrasts the nature of risk and uncertainty as ways of defining the type of judgments required in innovation management: "Risk has its place in a calculus of probabilities. It applies to a specific course of action. The risk of an action is the likelihood that it will produce an unwanted result. *Risk lends itself to quantitative expression*, as when we say that the chances of failing to strike oil in a field are better than fifty-fifty. . . . Uncertainty is quite another matter. *A situation is uncertain when it requires action but resists analysis of risks.* A gambler takes a risk in an honest game of blackjack when, knowing the odds, he calls for another card. The same gambler, unsure of the odds and the honesty of the game, is in a situation of uncertainty. He can act, but he cannot estimate the risks or rewards of his action. . . . An explorer lost in the woods, short of food and water, confronts even greater uncertainty: he must act even though relevant alternatives are undefined. He must *invent* what to do. He has no way of calculating with any precision the risks of action. He has only rough quidelines of *skill and experience* to help him" (*Technology and Change*, p. 21; my italics). Schon goes on to argue that many decisions about radically new technology are more aptly understood as uncertainty than as risk; see pp. 19–41 for the full argument.

69. The description of individual cases in which such a replacement has occurred does not constitute a definitive answer to the question. Noble, "Forces of Production," and Shaiken, "Automation," both argue against the mythology that de-skilling automated systems are the result of an inevitable replacement of skill with computer-embodied theory.

70. Glick, "Irrigation Canals," pp. 166–167.

71. Packer, "Ostia," pp. 360–361, 386–387.

72. M. Smith, "Hall and North," passim.

173. Rosenbloom, "Mechanization," passim. The two other explicit treatments of worker strife in TC are recent. D. Jones ("Tear Gas") shows fear of labor strife as a major cause of the popularity of tear gas in police departments during the early twenties, though without a study of worker motives for striking. Thomas ("Diesel") notes the Diesels'—father and son—technocratic concern about class warfare.

74. All the passages cited here are from Harris and Pris, "Memoirs," pp. 214–215.

75. Besides Noble and Shaiken (see note 69) the following works are significant. M. Smith's *Harpers Ferry* is a remarkably explicit analysis of the problem. Chapter

2 of Kasson's *Civilizing the Machine* calls attention to labor–management tensions at the Lowell textile mills. The presentation of four papers by Arthur Donovan, Daryl Hafter, Judith McGaw, and James Peterson at the 1979 annual meeting of SHOT, which study technology from the workers' perspective, are further evidence of new interest in the problem.

76. Edelstein, "Allerley Matkel," pp. 298–299.

77. Shelby, "Keystones," p. 548.

78. Hacker, "Greek Catapults," p. 50.

Chapter 4

1. I am indebted to conversations and correspondence with Melvin Kranzberg, Carl Condit, Thomas P. Hughes, and other early SHOT leaders for most of this information.

2. See Staudenmaier, *Design and Ambience*, appendix 14, for a more detailed discussion of the cultural references found in TC and, in particular, their individualistic character.

3. For a brief but helpful discussion of the diffusion and transfer models see Ruttan and Hayami, "Transfer," pp. 121–122. See chapter 2, pp. 54–55, for TC's references to diffusion of innovations.

4. Fox, "Fire Pistons," and Roland, "Bushnell."

5. Kren, "Chilinder," passim.

6. Feller, "Cotton Industry"; Rosenberg, "Transfer"; Ruttan and Hayami, "Transfer"; and Woodruff and Woodruff, "Diffusion." In all four cases the thematic interpretation is not primarily focused on vehicles of transfer but on problems related to the support network or to cultural tensions in transfer.

7. Ruttan and Hayami, "Transfer," pp. 124–125.

8. Rosenberg, "Transfer," pp. 558–563.

9. Ibid., p. 568.

10. Ibid., p. 570.

11. Ibid.

12. Dalrymple, "Tractors to Russia."

13. Of the nine author of theoretical models Jack Baranson, Arthur Goldschmidt, Yugiro Hayami, and Robert Theobald are economists; Nathan Rosenberg and William Woodruff are economic historians; and the remaining three are scholars in fields not typical of SHOT's author constituency. Peter F. Drucker is a specialist in management, Ritchie Calder is a political scientist, and Donald W. Shriver, Jr., is a philosopher.

The six case-study authors with economic expertise are Dana G. Dalrymple, Irwin Feller, J. Clayborn La Force, Gerhard Rosegger, Shannon R. Brown, and S. J. Nicholas.

14. Goldschmidt, "Emerging Nations," pp. 587–589.

15. Baranson, "Adapting Technology," p. 26.

16. Theobald, "Emerging Nations," pp. 606–607. For other articles calling for respect of the recipient cultures, see appendix 3.18.4.B.

17. Both quotes are from Shriver, "Man and Machines," pp. 550–551.

18. For seven articles referring to technology as part of colonial policy see appendix 3.18.2.E. In particular see Jensen and Rosegger, "Lower Danube," pp. 679–680 and passim. Finally, note my discussion of TC's typical avoidance of colonialism in chapter 5.

19. Fries, "British Response," esp. p. 403. Brown, "Ewo Filature," passim.

20. Hughes, "British Lag," p. 39.

21. Dorn, "Cooper and Détente," esp. p. 335.

22. Hacker, "China and Japan," p. 52. For a more thorough study of Japanese weaponry with an elegantly developed cultural theme, see Perrin, *Giving Up the Gun*.

23. Sinclair, "Canadian Technology," p. 109.

24. Jensen and Rosegger, "Lower Danube," p. 679.

25. Hacker, "China and Japan," p. 44; Hughes, "British Lag," p. 39; and Jensen and Rosegger, "Lower Danube," p. 680. For a recent monograph treating cultural tensions related to transfer in great detail, see M. R. Smith, *Harpers Ferry*. Finally see my discussion of TC's overall response to the cultural aspects of transfer in chapter 5.

26. This and the following two texts are fron Rürup, "History of Technology," p. 162.

27. Ibid., pp. 165–166.

28. Possible exceptions are Mumford, "Nature of Man," p. 307, and Daniels, "Big Questions," p. 3. Both argue that culture determines technological patterns.

29. Ellul, "Technological Order," pp. 394–395.

30. It is worth recalling our discussion of the systems approach to emerging technology in chapter 2 ("System as a Sector of the Economy"). The tendency to expand the systems model to include ambient factors as functional systemic components ultimately guts the historical perspective. Ellul makes the same point here.

31. Thus Morris argues that profit maximization is the key norm generating a deterministic technological force ("Context," p. 147).

32. Ellul, "Technological Order," p. 395.

33. For this and the following brief quotations see ibid., pp. 408ff.

34. Both quotations from Buchanan, "Exploitation," p. 543. See Clarke, "Technology and Man," pp. 423–424, for precisely the same point.

35. Ferguson, "Toward a Discipline," pp. 16–17. See Clarke, "Technology and Man," pp. 428–429 for the same point.

36. Mumford, "Neglected Clue," p. 231.

37. Heilbroner, "Machines," p. 336.

38. Ibid. For the following brief quotations, see pp. 337–338.

39. Ibid., p. 338.

40. Heilbroner asserts the fixed-sequence character of scientific knowledge, with some hesitancy, in his fifth footnote to the passage under discussion. Ibid., pp. 338–339.

41. Both quotations from ibid., pp. 343–344.

42. Ibid., p. 340.

43. For a presentation of the Marxist position see Zvorikine, "Technology," passim. For a perceptive reflection on the possibilities and disappointments of Marxist history of technology to the present, see Joravsky, "Marxism," pp. 6–7.

44. Ogburn, *Social Change*, passim. Allen, "Social Change," exemplifies the frame of reference that streses only the impact of technology on culture. He is an explicit advocate of Ogburn's cultural lag theory.

45. Rürup, "History of Technology," p. 188. Rürup proceeds immediately to note that Ogburn's theory, and determinist theories in general, are consonant with a twofold interpretation of technology and society wherein premodern and non-Western technologies are seen as governed by cultural norms whereas modern Western technology is autonomous: "It appears that one can quite comfortably proceed from the thesis that throughout the history of mankind technology has always developed in reaction to social needs—a thesis one can continue to maintain, *at the same time assuming that a different type of causal relationship applies to the present.* According to this way of thinking, *modern industrial technology* emerged under specific historical conditions and in response to needs one can assume existed; yet over the course of time *this technology became increasingly independent of such social conditions, to the extent that it is now obeying no laws but those of its own development.* Technology, to conclude this line of argument, *progresses on its own power after an initial takeoff phase*" (pp. 188–189; my italics).

46. Howland, "Engineering Education," pp. 4–5.

47. Price, "Technology Independent," p. 566. Although Price uses language that is deterministic in tone (see also his "Automata," p. 23), he is more subtle than Heilbroner. Price does not claim a neceesary link between all technology and prior scientific knowledge, nor does he assert a radical disjunction between pre- and postmodern technologies.

48. H. Jones, "Ideas," p. 25; L. Hunter, "Water-Mills," pp. 453–454; Post, "Page Locomotive," p. 142. See also my further discussion of failure studies in chapter 6.

49. Ferguson, "Toward a Discipline," p. 19.

50. Rürup, "History of Technology," pp. 174–175.

51. Woodruff and Woodruff, "Diffusion," p. 474. See also DeWalt, "Appropriate Technology."

52. Rürup, "History of Technology," p. 169.

53. Leibniz, *Monadology*.

54. See Lienhard, "Rate of Improvement," for his measures of the rapidly changing "rate of improvment" in the West since 1700.

55. See Rürup, "History of Technology," pp. 188–190, for a discussion of these dimensions of recent technology and their relationship with determinism. For other articles with explicit reference to the unforeseen consequences of technological change, see appendix 3.20.2.

56. Ferguson, "Toward a Discipline," pp. 23–24. The following summary of the Hughes model is taken from his original article, "Technological Momentum," published in *Past and Present* (August 1969).

57. Chapter 2's discussion of the goal-directed dimension of development projects is a prime instance of design flexibility in the early stages of emerging technology.

See chapter 2, "The Process of Development." See also chapter 5 for a further elaboration of the momentum model.

58. Hounshell, "Telephone," pp. 160–161.

59. Kevles, "Postwar Research," p. 41.

60. Ibid., pp. 33–34.

61. Ibid., pp. 46–47.

62. Bailes, "Soviet Aviation," p. 73.

63. Ibid., p. 55.

64. Ibid., p. 75.

65. Ibid., p. 79.

66. The attentive reader will note that I am shifting roles in this treatment of momentum and, indeed, throughout chapter 4. While I am attempting to record faithfully the historiographical status of the themes—transfer, determinism, and momentum—in TC's current discourse, I am also beginning to argue for a set of interpretations of technological change which are, in my judgment, critically important for the history of technology in the future. Thus my critique of determinism is considerably more explicit than that of any TC historians except Ferguson and Rürup, and my stress on the cultural question in transfer, together with my assertion of the importance of the momentum model, go beyond a neutral reporting of the state of TC's discourse. The full articulation of my position can be found in chapter 5.

67. Bryant, "Thermodynamics," pp. 152–153.

68. Hounshell, "Telephone," p. 157. Correspondence is from Elisha Gray to W. D. Baldwin, 1 November 1876, and to A. L. Hayes, 2 November 1876.

69. L. Hunter, "Water-Mills," pp. 453–454.

70. Burke, "Bursting Boilers."

71. Chapin, "Patent Interferences"; Feller, "Cotton Industry"; Jeremy, "Textiles."

72. T. Smith, "Whirlwind," p. 453.

73. For references in historiographical and nonhistorical essays to the impact of cultural values on technical design, see Daniels, "Big Questions," pp. 6ff.; Durbin, "Values," p. 558; Mumford, "Nature of Man," passim; Shriver, "Man and Machines," pp. 532ff.; and Skolimowski, "Thinking," p. 382.

74. D. Jones, "Tear Gas."

75. Esper, "English Longbow." See also Tascher, "Cleveland," for another example.

76. Jensen and Rosegger, "Lower Danube," pp. 686, 688, 692, 695, 696; DeWalt, "Appropriate Technology," p. 32.

77. Pursell, "Government," pp. 173–174; Susskind and Inouye, "Technology Assessment," passim.

Chapter 5

1. Mumford, "Authoritarian Technics," pp. 6, 8.

2. For example: "Lewis Mumford's *Technics and Civilization* had such a great impact on me; I read it as an undergraduate at Amherst when it was first published back

in 1934, and it has stuck with me ever since—and I continue to assign it to my classes because it is still a mind-blowing experience for them!" (Kranzberg to Staudenmaier, 4 March 1983).

3. The critique of the ideology of progress I propose below is perhaps unorthodox in emphasis. Nannerl O. Keohane's excellent survey of the origins and the literature of the idea of progress, for example, suggests that Descartes and Bacon provided the catalyst for gathering a set of earlier Greek and Judeo-Christian ideas into a progressive ideology ("Enlightenment Idea of Progress"). I agree and, indeed, I agree with her focus on the critical link between knowledge and control over nature in their thinking. We diverge when Keohane identifies their commitment to the radically new scientific method with "vigorous optimism" (p. 29). For me the key to the Cartesian and Baconian contribution is not optimism but anxiety. Descartes's obsession with certitude leads directly to his famous disjunction of valid method from human context. It is this split—method from context—which, even more than the linking of knowledge and power, creates the conceptual core of the ideology of progress. Put another way, the obsession with certitude grounds the correlative obsession with control that marks Cartesian and Baconian method.

It will not surprise the reader that I stress the split of method from context in my critique of the ideology of autonomous progress. My analysis of the applied science and technological determinism debates in TC has convinced me that it is the insistence on preserving scientific and technological praxis from the "contamination" of contextual factors that has given the myth of autonomous Western progress its social, political, and economic power. This same method-context disjunction militates against any genuine contextual approach to the history of technology. The following brief analysis attempts to clarify these links.

Langdon Winner's *Autonomous Technology* probes Bacon's method-context disjunction from another perspective by stressing Bacon's distrust of politics and his desire to preserve the pure motivation of scientific control over nature. See pp. 21–25 and 135–139.

4. Bunge, "Applied Science," p. 330. See Laudan for a somewhat more implicit and subtle parallel position (*Progress and Its Problems*, pp. 101, 131). For an insightful analysis of the Cartesian assumption as it has influenced Western science see Keller and Grontkowski, "Mind's Eye," pp. 212, 214. Keller and Grontkowski situate the Cartesian disjunction in the much longer Western tradition of Greek philosophy. For a portrayal of Greek ambivalence about scientific and technological progress in the dramatic poetry of Aeschylus, Sophocles, and Euripides see Kahn, "Greek Tragedies."

For a similar interpretation of Descartes see Hans-Georg Gadamer, *Truth and Method*, pp. 245–253. Robert Nisbet makes the same point in his somewhat cavalier treatment of Descartes (*Idea of Progress*, pp. 115–117). For one of Michael Polanyi's frequent passing references to the disjunction see *Personal Knowledge* (1958), p. 269. Finally, for a thorough analysis of the relationship between the "cogito" and Cartesian theory of scientific knowledge see James Collins, *Descartes*.

5. Heilbroner, "Machines," p. 336.

6. The following passage from Feibleman is typical of the style. "Any practice that adheres too strongly to tradition is sure to be conservative and in this way *to stand in the way of improvement*. Thus tradition, which once was held to be the one indispensable ingredient in any article made by man, now is regarded merely as a

primitive skill which may *stand in the way* of a more efficient one" ("Skills," pp. 323–324; my italics).

7. One text suffices to capture the Baconian flavor. "I come in very truth leading to you Nature with all her children to bind her to your service and make her your slave." For a careful study of the Baconian rhetoric of masculine control in science and feminine exploitation in nature, see Evelyn Fox Keller, "Baconian Science." The text is found on page 301.

8. See my discussion of the common link between systems theory and the commitment in industrial capitalism to hierarchical control and standardization rather than negotiation and kindship (chapter 2, "System as a Sector of the Economy"). For a discussion of the congruence between the ideology of progress and the invisible hand model, see Richard E. Sclove, "Energy Policy," pp. 52–53.

9. The text is cited in Lowell Tozer's discussion of the Hall of Science. In his description of the iconography of the exposition he stresses its ideology of human passivity and technological progress: "In the Hall of Science at the Chicago Century of Progress International Exposition in 1933 stood a large sculptural group containing a nearly life-sized man and woman, hands outstretched as if in fear or ignorance. Between them stood a huge angular robot nearly twice their size, bending low over them, with an angular metallic arm thrown reassuringly around each" ("Century of Progress," p. 206). Carroll W. Pursell called my attention to Tozer ("Government," p. 162).

10. Langdon Winner's seminal *Autonomous Technology* and his forthcoming *Paths in Technopolis: Essays on Technology, Politics, and the Environment* are the most thorough and thoughtful exploration of the myth of progress. For a shorter and elegant treatment of the myth, covering its philosophical content, its religious and cultural roots, and much of twentieth-century comment on it, see Keohane, "Enlightenment Idea of Progress," passim and esp. pp. 32–40.

11. Daniels, "Big Questions," p. 2. See also, chapter 4, p. 147.

12. Porter, "Historiographical Revolution," p. 2.

13. Ibid. Kranzberg is aware of the same pattern. "You [Staudenmaier] note that 'the field is largely ignored by mainstream historians.' In a sense, that is true, and that is why we were so happy when Roe Smith was awarded a prize by the OAH for his book, for it marked the recognition by the larger field of American historians that a book on the history of technology was truly a contribution. That is why we were also happy when Dave Hounshell's piece on the history of technology appeared in the *Journal of American History*" (Kranzberg to Staudenmaier, 1 April 1983, p. 6).

E. P. Thompson critiques this split between historical specialization and the larger culture in highly similar terms for labor historians. "There has never been any single type of 'the transition' [from pre-industrial to industrial patterns]. The stress of the transition falls upon *the whole culture*: resistance to change and assent to change includes the systems of power, property relations, religious institution, etc., inattention to which merely *flattens phenomena and trivializes analysis*" ("Time," p. 80; my italics).

In a similar vein Herbert Gutman critiques traditional American labor history. "Its methods encouraged labor historians *to spin a cocoon around American workers, isolating them from their own particular subcultures, isolating them from the larger national culture.* An increasingly narrow 'economic' analysis caused the study of American working-class history to grow more constricted and become more detached from larger developments in American social and cultural history and from the writing of American social and cutural history itself" (*Work*, pp. 9–11; my italics).

14. The claim that technology is value neutral and the correlative critique of those who dare to speak against "progress" is embodied in the following passage from a full-page United Technologies advertisement from 1983. Note the striking use of the word "technology" in the singular. In the rhetoric of progress no distinctions can be made among various technologies; all are part of a singular moving force. "Ethically, technology is neutral. There is nothing inherently either good or bad about it. It is simply a tool, a servant, to be be refined, directed and deployed by people for whatever purposes they want fulfilled. . . . So fast do times change, because of technology, that *some people, disoriented by the pace, express yearning for simpler times. They'd like to turn back the technological clock.* But longing for the primitive is utter folly" (my italics). In other words, to criticize any technical design is to regress; therefore all design decisions are progress. I am indebted to Colleen Dunlavy of the MIT Science, Technology, and Society Program for the advertisement.

15. For our purposes the common expression "Whig history" can be defined as those interpretations of technological change that foster the ideology of autonomous technological progress by assuming that technological method and design can be interpreted in isolation from their ambience.

D. R. Oldroyd's recent reflections on Sir Archibald Geikie and Whig historiography of science ("Whig Historiography") raises the question of whether historians of science can critique earlier historiography of science without falling victim to their own brand of Whiggishness. By applying present-day norms of historiography to earlier historians, Oldroyd argues, historians of science stand guilty of appraising past historiographical events "in terms of present circumstances and values, rather than in terms of their own time" (p. 444).

By thus centering the definition of Whig history in the sin of anachronism, Oldroyd rightly alerts historians to the limitations of their own historiographical assumptions (see chapter 1, "Historical Scholarship within a Cognitive World View"). My own use of the term, however, centers the definition of Whig history more in the philosophical commitment to sever method from context than in the related tendency to anachronistic success stories. Thus, while I agree with Oldroyd's cautions about being Whiggish about Whig historiography and find his suggested resolution quite helpful, it is not central to my critique. My use of the expression should not be assumed to be identical with its common usage as defined by either Butterfield ("Whig Interpretation") or Oldroyd.

16. Kranzberg to Staudenmaier, 1 April 1983, pp. 3–4.

17. Mayr, "Historiographic Problem," p. 664.

18. See chapter 2, "Innovation." Billie DeWalt's study of the Mexican seed drill ("Appropriate Technology") comes the closest to a situation where distinct cultural perspectives influence the acceptance of a technological design.

19. See Staudenmaier, *Design and Ambience*, pp. 83–84.

20. In fact a reasonable balance exists between internalist and externalist articles. While contextual articles hold the predominant share with 53 percent, internalist articles (17 percent) are only slightly more common than externalist (14 percent). During the most recent seven-year period both styles are represented by fourteen articles (15.5 percent each). See appendix 2.

21. On technological support networks and technical traditions see the appropriate sections of chapter 2.

In my discussion of the limitations of the systems model (chapter 2), I note that the popularity of the systems approach has one significant liability along with its virtues. When systems thinking becomes the dominant interpretation of technological change, especially when it is applied to large-scale economic models, the distinction between internal design symmetry and its nonsystemic ambience tends to be lost. As the systems model is applied at larger and larger levels of abstraction, the ambivalent and ambiguous dimensions of the historical context tend to be treated as if they were functional components rather than as independent historical variables. Clearly, if the systems model were to be accepted as the dominant interpretative model for emerging technology it would foster the myth of progress by its seeming ability to convert ambient factors into design components.

22. Ferguson, "Toward a Discipline," p. 19. See "The Technological Ambience" above in this chapter and Louis Hunter's call for study of "the absence of change" in chapter 4, "The Enduring Nature of Existing Technological Artifacts."

23. Woodbury, "Whitney Myth." In this discussion the term "successful" is not meant to be a value judgment about the contribution of a given technical design to human welfare. The expression refers only to the technology's eventual adoption—for good or ill—by some human culture.

24. Leslie, "Cooper-Cooled Engine."

25. Thomas, "Diesel"; Jensen and Rosegger, "Lower Danube"; Cowan, "Women and Technology."

26. Rürup, "History of Technology," pp. 191–192; Lieberstein, "Soviet NOT Movement"; Rosenbloom, "Mechanization"; and Mumford, "Authoritarian Technics" and "Nature of Man."

27. Jones, "Tear Gas," and Wilkinson, "Anticipating Taylor."

28. Seventy-six articles treat transfer. The seven historical studies, together with the five nonhistorical essays containing explicit treatments of cultural dimensions, constitute 15.7 percent of the total. See chapter 4, "Transfer and Culture," for discussions of the seven historical articles. For the five nonhistorical essays see appendix 3.18.4.B.

29. Dorn, "Cooper and Détente," p. 337.

30. Jensen and Rosegger, "Lower Danube"; Sinclair, "Canadian Technology"; Brown, "Ewo Filature"; Hacker, "China and Japan."

31. Gorman, "Telegraph in India."

32. Jensen and Rosegger, "Lower Danube." Again, Baranson, "Adapting Technology"; Calder, "Technology in Focus"; Drucker, "Ancient Jobs"; Shriver, "Man and Machines"; and Theobald, "Emerging Nations" raise the question in nonhistorical essays. Woodruff and Woodruff address the same issue in their essay on the unrepeatable character of Western development ("Diffusion").

The two studies of appropriate technology are Hacker, "China and Japan," and DeWalt, "Appropriate Technology"

33. The monumental work of Joseph Needham and the more recent research of Nathan Sivin and other orientalists (see, for example, *Science and Technology in East Asia*, 1977) does not yet appear to have had much influence in the mainstream of SHOT research.

34. Rürup, "History of Technology," pp. 180–181; Rosenbloom, "Mechanization," pp. 499–502; Durbin, "Values," pp. 557–558; Zvorikine, "Soviet History" and "Tech-

nology," passim; Heilbroner, "Machines," p. 345; Joravsky, "Marxism," passim. Note that Rürup parallels his critique in the passage cited in this note.

35. Mayr, "Adam Smith"; Thomas, "Diesel"; and Dorn, "Cooper and Détente." John Burke's two case studies ("Bursting Boilers" and "Advertising") focus some attention on U.S. governmental policies as influenced by private enterprise. In addition, Arthur D. Kahn ("Greek Tragedies") and all the articles of Lynn White, Jr., treat questions that could also be considered ancillary to an exploration of Western capitalism. White raises the question of the influence of Roman Christianity on the West's style of technology—as exploitation of, rather than harmony with, nature. Kahn presents textual evidence of a serious critique by Greek tragedians of the value of an aggressive domination of nature. For a recent critique of precisely the same sort see Carolyn Merchant's *Death of Nature* (1980) or her essay "Mining the Earth's Womb," in Rothschild, ed., *Machina ex Dea* (1983).

The relationship between capitalism and the nineteenth- and twentieth-century adoption of hierarchical, systemic standardization in place of a technological style that demands negotiation has evoked little comment in TC articles. For my discussion contrasting TC's references to technical traditions (centered in negotiation) and to systems (centered in hierarchical control), see chapter 2.

36. Cowan, "Revolution" and "Women and Technology." The emergence of SHOT's WITH subcommittee in 1976 (TC 18 [July 1977]: 496) begins to show signs of convergence with the work of such scholars as historian of science Evelyn Fox Keller (see note 7 above) and developmental psychologist Carol Gilligan (*Different Voice*). Keller and Gilligan have convincingly demonstrated that "masculine" stereotypes have exerted a dominant and distorting influence in choices of scientific interpretative models (Keller) and in all of the major developmental psychological models (Gilligan). Joann Vanek's "Housework" and Charles A. Thrall's "The Conservative Use of Modern Household Technology" (TC 23 [April 1982]) are at this time the only other women's studies published in the journal.

37. See my forthcoming review essay for TC of Rothschild's *Machina ex Dea.*

38. See the beginning of chapter 1, and appendix 2 in Staudenmaier, *Design and Ambience.*

39. See chapter 2, "The Linguistic Origins of The 'Invention-Development-Innovation' Model." See also Rürup, "History of Technology," p. 183, and Hughes, "Themes," pp. 703, 705–706, for their calls to adopt economic perspectives for the history of technology.

40. TC 22 (July 1981), Dexter Prize citation, p. 562

41. TC 23 (July 1982), Dexter Prize citation, p. 442.

42. TC 24 (July 1983), Dexter Prize citation, p. 468.

43. The Dexter Prize citation for 1983 was provided before publication in TC 25 (July 1984) by Joan Mentzer at the journal's editorial office.

44. Thackray, "Franklin Institute: Review," p. 384.

45. Glenn Porter, "Commentary," pp. 2–3. For my earlier observations of the imbalance toward origin studies in TC see chapter 2, "An Overview of the Theme." For a discussion of the importance of TC's "success-story" pattern see chapter 5, "The Roads Not Taken" and chapter 4, "The Debate over Technological Determinism: The Second Corollary."

46. Wallace, perhaps due to his relatively recent focus on technological designs, is somewhat less successful in integrating the community dynamics of Rockdale with the design of its technologies. For this reason I have stressed Smith's integration in the text.

Stuart Leslie notes the importance of the small-community focus for the success of Smith and Wallace together with that of Hunter's *Waterpower* in "integrating technology and social environment" in his review of an exhibition at the Milwaukee Public Museum ("Urban Habitat," pp. 427–428).

For one very recent TC article that adopts a community study matrix see DeWalt, "Appropriate Technology."

47. It is worth noting as we pass from community studies to worker history that a significant group of community studies has appeared in the past two decades as part of the mainstream of U.S. labor history. However, with the exception of Aitken, M. R. Smith, and the three works discussed below (Brody, Meyer, and Dublin) the critical integration of the technological design of the production process and the workers' perspective is not treated with the same sophistication.

Some of the more significant studies are Bodnar, *Immigration and Industrialization* (1977); Buder, *Pullman* (1967); Byington, *Homestead* (1969); Cumbler, *Working-Class Community* (1979); Dawley, *Class and Community* (1976; Edwards, *Contested Terrain* (1979); Foster *Class Struggle* (1974); Gutman, *Work, Culture, and Society* (1976); Hareven and Langenbach, *Amoskeag* (1978); Hirsch, *Crafts in Newark* (1978); G. S. Jones, *Outcast London* (1971); Montgomery, *Workers' Control in America* (1979); Nelson, *Managers and Workers* (1975); Prude, *Industrial Order* (1983); Stearns and Walkowitz, eds., *Workers in the Industrial Revolution* (1974); Walkowitz, *Worker City, Company Town* (1978); and Weber, *Warren, Pennsylvania* (1976).

48. On TC avoidance of a worker perspective see chapter 3, "Technical Skill," and chapter 5, "Questions Seldom Asked," pp. 288–290. Rürup's pessimistic view of labor history is further evidence of the near absence of this perspective in the field before 1974 ("History of Technology," pp. 191–192).

49. Aitken, *Taylorism* (1960).

50. Noble, *America by Design* (1979) and *Forces of Production* (1983). The earliest presenters at SHOT meetings, Daryl Hafter (1976), Susan Levine (1976), and Caroline Fuller Sloat (1978), combined an interest in workers and women. The first full session on technology and workers in 1978 featured papers by David Noble, Richard Horowitz, and Randolph Langenbach. It was followed by another session in 1979 (Arthur Donovan, Daryl Hafter, Judith McGaw, and James Peterson). See "Organizational Notes" in TC 18 (July 1977): 485–486, TC 20 (July 1979): 602–604, and TC 21 (July 1980): 460–462.

51. Gutman, *Work, Culture, and Society* pp. 3–19.

52. Chandler, *Visible Hand* (1977); Jenkins, *Images and Enterprise* (1975); Layton, *Revolt* (1971).

SHOT recognized the importance of Alfred Chandler's work linking technological development to institutional history in its 1978 annual meeting (TC 20 [July 1979]: 598) where a session was held to discuss *The Visible Hand* (1977). Thomas P. Hughes's *Networks of Power* (1983) is another very recent work that makes extensive use of the institutional perspective.

Not surprisingly, I would like to see more of a critique of capitalism included in Jenkins's and Chandler's otherwise remarkably thorough discussions of business institutions.

53. Koppes, *JPL* (1982); Roland, *Model Research* (1984); Allison, *New Eye* (1981); Mack, *Landsat* (1983); and M. R. Smith, ed., *Military Enterprise* (forthcoming).

54. Hughes, *Networks*, p. 6.

55. Trescott, ed., *Dynamos and Virgins* (1979); Rothschild, ed., *Machina ex Dea* (1983); D'Onofrio-Flores and Pfafflin, eds., *Role of Women* (1982); and Cowan, *More Work for Mother* (1983).

56. For a discussion of many of the dynamics to be treated in this model from a cybernetic perspective, see Zwick, "Dialectical Thermodynamics," in Lasker, ed., *Applied Systems and Cybernetics* (1981), pp. 618–626.

57. For a discussion of this point recall Donald Schon's enlightening distinction between risk and uncertainty thinking in corporate handling of emerging technology. (chapter 3, note 68).

58. See Hounshell, *American System*, for the most thorough study of the machine-tool tradition.

59. For a thoughtful presentation of the world view at work in Ford's assembly-line design stage see Meyer, *Five Dollar Day*, passim.

60. This is, of course, the situation that led Thomas Hughes to articulate the momentum model. The design constituency of the hydrogenation technology in Germany maintained the technology in the face of Weimar Germany's relatively inhospitable climate and eventually threw its lot in with the Nazis because of the promise of further support for "their" technology. For another case see T. Smith, "Whirlwind" (chapter 2, "The Process of Development").

61. The Ford mass-production design was modified but not radically changed by the Sloan marketing system of the twenties.

62. The tendency of members of the impact constituency to repress their awareness of what and how they lose has been articulated with great elegance in Paulo Freire's *Pedagogy of the Oppressed* (1970). Others who attend to the same dynamics include Zuboff, "Work and Human Interaction" (pp. 1–5), and Noble, "Luddites."

63. This articulation is of course oversimple. Social Darwinism and other similar theories would suggest that the personal tragedies of the impact constituency are a necessary "survival of the fittest" process by which society excretes its unfit members.

64. Perhaps the best studies of cultural conflict due to imposed technological change have emerged in those labor history studies that consider the beginning of capitalistic industrialization. E. P. Thompson (*English Working Class*) and Herbert Gutman (*Work, Culture, and Society*) come immediately to mind, as does M. R. Smith's *Harpers Ferry*. Shoshanah Zuboff's unpublished manuscript "Work and Human Interaction" is an elegant synthesis of much of this scholarship.

65. The synthetic use of the contextual approach to interpret the technological style of an entire culture is, of course, much more complex than these few lines can indicate. My own attempt at such an interpretation of the United States' twentieth-century style is currently in process.

Bibliography

Articles in Technology and Culture *(with short titles)*

1. Agassi, Joseph. The confusion between science and technology in the standard philosophies of science. Vol. 8, no. 3 (Summer 1966): 348–366. ("Confusion")

2. Allen, F. R. Technology and social change: Current status and outlook. Vol. 1, no. 1 (Winter 1959): 48–59. ("Social Change")

3. Bachrach, Bernard S. The origin of American chivalry. Vol. 10, no. 2 (April 1969): 166–171. ("American Chivalry")

4. Bachrach, B.S. Early medieval fortifications in the "West" of France: A revised technical vocabulary. Vol. 16, no. 4 (October 1975): 531–569. ("Fortifications")

5. Bachrach, B. S. Fortifications and military tactics: Fulk Nerra's strongholds circa 1000. Vol. 20, no. 3 (July 1979): 531–549. ("Fulk Nerra")

6. Bailes, K. E. Technology and legitimacy: Soviet aviation and Stalinism in the 1930's. Vol. 17, no. 1 (January 1976): 55–81. ("Soviet Aviation")

7. Balmer, R. T. The operation of sand clocks and their medieval development. Vol. 19, no. 4 (October 1978): 615–632. ("Sand Clocks")

8. Baranson, Jack. Economic and social considerations in adapting technologies for developing countries. Vol. 4, no. 1 (Winter 1963): 22–29. ("Adapting Technology")

9. Bedini, Silvio A. The compartmented cylindrical clepsydra. Vol. 3, no. 2 (Spring 1962): 115–141. ("Clepsydra")

10. Bedini, S. A. The role of automata in the history of technology. Vol. 5, no. 1 (Winter 1964): 24–42. ("Automata")

11. Bedini, S. A. The evolution of science museums. Vol. 6, no. 1 (Winter 1965): 1–29. ("Museums")

12. Bilstein, Roger E. Technology and commerce: Aviation in the conduct of American business, 1918–1929. Vol. 10, no. 3 (July 1969): 392–411. ("Aviation")

13. Bowles, Edmund A. On the origin of the keyboard mechanism in the late Middle Ages. Vol. 8, no. 2 (Spring 1966): 152–162. ("Keyboard")

14. Boyer, Marjorie Nice. Medieval pivoted axles. Vol. 1, no. 2 (Winter 1959): 128–138. ("Axles")

15. Boyer, M. N. Rebuilding the bridge at Albi, 1408–1410. Vol. 7, no. 1 (Winter 1966): 24–37. ("Bridge")

16. Braun, Wernher von. The Redstone, Jupiter, and Juno. Vol. 4, no. 4 (Fall 1963): 452–465. ("Redstone")

17. Bray, Francesca. Swords into plowshares: A study of agricultural technology and society in early China. Vol. 19, no. 1 (January 1978): 1–31. ("Early China")

18. Brittain, James E. The introduction of the loading coil: George A. Campbell and Michael I. Pupin. Vol. 11, no. 1 (January 1970): 36–57. ("Loading Coil")

19. Brittain, J. E., and Robert C. McMath. Engineers and the New South creed: The formation and early development of Georgia Tech. Vol. 18, no. 2 (April 1977): 175–201. ("Georgia Tech")

20. Brown, Shannon R. The Ewo filature: A study in the transfer of technology to China in the 19th century. Vol. 20, no. 3 (July 1979): 550–568. ("Ewo Filature")

21. Bryant, Lynwood. The silent Otto. Vol. 7, no. 2 (Spring 1966): 184–200. ("Silent Otto")

22. Bryant, L. The origin of the four-stroke cycle. Vol. 8, no. 2 (April 1967): 178–198. ("Four-Stroke Cycle")

23. Bryant, L. The role of thermodynamics in the evolution of heat engines. Vol. 14, no. 2 (April 1973): 152–165. ("Thermodynamics")

24. Bryant, L. The development of the diesel engine. Vol. 17, no. 3 (July 1976): 432–446. ("Diesel")

25. Buchanan, Scott. Technology as a system of exploitation. Vol. 3, no. 4 (Fall 1962): 535–543. ("Exploitation")

26. Bunge, Mario. Technology as applied science. Vol. 7, no. 3 (Summer 1966): 329–347. ("Applied Science")

27. Burke, John G. Bursting boilers and the federal power. Vol. 7, no. 1 (Winter 1966): 1–23. ("Bursting Boilers")

28. Burke, J. G. Wood pulp, water pollution, and advertising. Vol. 20, no. 1 (January 1979): 175–195. ("Advertising")

29. Burlingame, Roger. The hardware of culture. Vol. 1, no. 1 (Winter 1959): 11–19. ("Hardware")

30. Burlingame, R. Technology: Neglected clue to historical change. Vol. 2, no. 3 (Summer 1961): 219–229. ("Neglected Clue")

31. Burns, Alfred. Ancient Greek water supply and city planning: A study of Syracuse and Acragas. Vol. 15, no. 3 (July 1974): 389–412. ("Greek Water Supply")

32. Cain, Louis P. Raising and watering a city: Ellis Sylvester Chesbrough and Chicago's first sanitation system. Vol. 13, no. 3 (July 1972): 353–372. ("Chicago Sanitation")

33. Cain, L. P. Unfouling the public's nest: Chicago's sanitary diversion of Lake Michigan water. Vol. 15, no. 4 (October 1974): 594–613. ("Watering a City")

34. Calder, Ritchie. Technology in focus. Vol. 3, no. 4 (Fall 1962): 563–580. ("Technology in Focus")

35. Camp, L. Sprague de. Master gunner Apollonios. Vol. 2, no. 3 (Summer 1961): 240–244. ("Apollonios")

36. Cardwell, Donald S. L. Power technologies and the advance of science, 1700–1825. Vol. 6, no. 2 (Spring 1965): 188–207. ("Power Technology")

37. Cardwell, D. S. L. The work of James Prescott Joule. Vol. 17, no. 4 (October 1976): 674–687. ("Joule")

38. Carver, J. Scott. *Tekhnologicheskii zhurnal*: An early Russian technoeconomic periodical. Vol. 18, no. 4 (October 1977): 622–643. ("Russian Journal")

39. Chapin, Seymour L. Patent interferences and the history of technology: A high-flying example. Vol. 12, no. 3 (July 1971): 414–446. ("Patent Interferences")

40. Clark, Charles J. The development of the semi-automatic freight-car coupler, 1863–1893. Vol. 13, no. 2 (April 1972): 170–208. ("Car Coupler")

41. Clarke, W. Norris, S.J. Technology and man: A Christian vision. Vol. 3, no. 4 (Fall 1962): 422–442. ("Technology and Man")

42. Claxton, Robert H. Miguel Rivera Maestre: Guatemalan scientist-engineer. Vol. 14, no. 3 (July 1973): 384–403. ("Maestre")

43. Condit, Carl. Sullivan's skyscrapers as the expression of nineteenth century technology. Vol. 1, no. 1 (Winter 1959): 78–93. ("Skyscrapers")

44. Condit, C. The structural system of Adler and Sullivan's Garrick Theater building. Vol. 5, no. 4 (Fall 1964): 523–540. ("Garrick Theater")

45. Condit, C. The first reinforced-concrete skyscraper: The Ingalls building in Cincinnati and its place in structural history. Vol. 9, no. 1 (January 1968): 1–33. ("Ingalls Building")

46. Constant, Edward W., II. A model for technological change applied to the turbojet revolution. Vol. 14, no. 4 (October 1973): 553–572. ("Turbojet")

47. Cowan, Ruth Schwartz. The "industrial revolution" in the home: Household technology and social change in the 20th century. Vol. 17, no. 1 (January 1976): 1–23. ("Revolution")

48. Cowan, R. S. From Virginia Dare to Virginia Slims: Women and technology in American life. Vol. 20, no. 1 (January 1979): 51–63. ("Women and Technology")

49. Dalrymple, Dana. The American tractor comes to Soviet agriculture: The transfer of technology. Vol. 5, no. 2 (Spring 1964): 191–214. ("Tractors to Russia")

50. Dalrymple, D. The development of an agricultural technology: Controlled-atmosphere storage of fruit. Vol. 10, no. 1 (January 1969): 35–48. ("Storage of Fruit")

51. Daniels, George H. The big questions in the history of American technology. Vol. 11, no. 1 (January 1970): 1–35. ("Big Questions")

52. Davison, C. St. C. Transporting sixty-ton statues in early Assyria and Egypt. Vol. 2, no. 1 (Winter 1961): 11–16. ("Sixty-Ton Statues")

53. DeWalt, Billie R. Appropriate technology in rural Mexico: Antecedents and consequences of an indigenous peasant innovation. Vol. 19, no. 1 (January 1978): 32–52. ("Appropriate Technology")

54. Dorn, Harold. Hugh Lincoln Cooper and the first détente. Vol. 20, no. 2 (April 1979): 322–347. ("Cooper and Détente")

55. Dornberger, Walter R. The German V-2. Vol. 4, no. 4 (Fall 1963): 393–409. ("German V-2")

56. Dresbeck, LeRoy J. The ski: Its history and historiography. Vol. 8, no. 4 (October 1967): 467–479. ("The Ski")

57. Drucker, Peter F. Work and tools. Vol. 1, no. 1 (Winter 1959): 28–37. ("Work and Tools")

58. Drucker, P. F. The technological revolution: Notes on the relationship of technology, science, and culture. Vol. 2, no. 4 (Fall 1961): 342–351. ("Technological Revolution")

59. Drucker, P. F. Modern technology and ancient jobs. Vol. 4, no. 3 (Summer 1963): 227–281. ("Ancient Jobs")

60. Dryden, Hugh L. Future exploration and utilization of outer space. Vol. 2, no. 2 (Spring 1961): 112–126. ("Outer Space")

61. Durbin, Paul T. Technology and values: A philosopher's perspective. Vol. 13, no. 4 (October 1972): 556–576. ("Values")

62. Edelstein, Sidney M. The Allerley Matkel (1532): Facsimile text, translation, and critical study of the earliest printed book on spot removing and dyeing. Vol. 5, no. 3 (Summer 1964): 297–321. ("Allerley Matkel")

63. Ellul, Jacques. The technological order. Vol. 3, no. 4 (Fall 1962): 394–421. ("Technological Order")

64. Emmerson, George S. L. T. C. Rolt and the Great Eastern affair of Brunel versus Scott Russell. Vol. 21, no. 4 (October 1980): 553–569. ("Great Eastern")

65. Esper, Thomas. The replacement of the longbow by firearms in the English army. Vol. 6, no. 3 (Summer 1965): 382–393. ("English Longbow")

66. Feibleman, James K. Pure science, applied science, technology, engineering: An attempt at definitions. Vol. 2, no. 4 (Fall 1961): 305–317. ("Pure Science")

67. Feibleman, J. K. Technology as skills. Vol. 7, no. 3 (Summer 1966): 318–328. ("Skills")

68. Feller, Irwin. The diffusion and location of technological change in the American cotton-textile industry, 1890–1970. Vol. 15, no. 4 (October 1974): 569–593. ("Cotton Industry")

69. Ferguson, Eugene S. Technical museums and international exhibitions. Vol. 6, no. 1 (Winter 1965): 30–46. ("Museums")

70. Ferguson, E. S. Toward a discipline of the history of technology. Vol. 15, no. 1 (January 1974): 13–30. ("Toward a Discipline")

71. Ferguson, E. S. The American-ness of American technology. Vol. 20, no. 1 (January 1979): 3–24 ("American-ness")

72. Finch, James Kip. Engineering and science: A historical review and appraisal. Vol. 2, no. 4 (Fall 1961): 318–332. ("Engineering and Science")

73. Finn, Bernard S. The science museum today. Vol. 6, no. 1 (Winter 1965): 74–82. ("Museums")

74. Flick, Carlos. The movement for smoke abatement in 19th-century England. Vol. 21, no. 1 (January 1980): 29–50. ("Smoke Abatement")

75. Fox, Robert. The fire piston and its origins in Europe. Vol. 10, no. 3 (July 1969): 355–370. ("Fire Piston")

76. Frankel, J. P. The origin of the Indonesian "Pamor." Vol. 4, no. 1 (Winter 1963): 14–21. ("Indonesian Pamor")

77. Frazier, Arthur H. Daniel Farrand Henry's cup-type "telegraphic" river current meter. Vol. 5, no. 4 (Fall 1964): 541–565. ("Current Meter")

78. Fries, Russell I. British response to the American System: The case of the small-arms industry after 1869. Vol. 16, no. 3 (July 1975): 377–403. ("British Response")

79. Fryer, David M., and John C. Marshall. The motives of Jacques de Vaucanson. Vol. 20, no. 2 (April 1979): 257–269. ("De Vaucanson")

80. Fullmer, June Z. Technology, chemistry, and the law in early 19th-century England. Vol. 21, no. 1 (January 1980): 1–28. ("Chemistry")

81. Russell, G. E. Farming systems of the classical era. Vol. 8, no. 1 (January 1967): 16–44 ("Farm Systems")

82. Gade, Daniel W. Grist milling with the horizontal waterwheel in the central Andes. Vol. 12, no. 1 (January 1971): 43–51. ("Grist Mills")

83. Garcia-Diego, J. A. The chapter on weirs in the Codex of Juanelo Turriano: A question of authorship. Vol. 17, no. 2 (April 1976): 217–234. ("Weirs")

84. Geise, John. What is a railway? Vol. 1, no. 1 (Winter 1959): 68–77. ("Railway")

85. Gilfillan, S. C. An attempt to measure the rise of American inventing and the decline of patenting. Vol. 1, no. 3 (Summer 1960): 201–213. ("Patents")

86. Glick, Thomas F. Levels and levelers surveying irrigation in medieval Valencia. Vol. 9, no. 2 (April 1968): 165–180. ("Irrigation Canals")

87. Goldschmidt, Arthur. Technology in focus: The emerging nations. Vol. 3, no. 4 (Fall 1962): 581–600. ("Emerging Nations")

88. Gorman, Mel. Sir William O'Shaughnessy, Lord Dalhousie, and the establishment of the telegraph system in India. Vol. 12, no. 4 (October 1971): 581–601. ("Telegraph in India")

89. Hacker, Barton C. Greek catapults and catapult technology: Science, technology and war in the ancient world. Vol. 9, no. 1 (January 1968): 34–50. ("Greek Catapults")

90. Hacker, B. C. The idea of rendezvous: From space station to orbital operations in space-travel thought, 1895–1951. Vol. 15, no. 3 (July 1974): 373–388. ("Rendezvous")

91. Hacker, B. C. The weapons of the west: Military technology and modernization in 19th century China and Japan. Vol. 18, no. 1 (January 1977): 43–55. ("China and Japan")

92. Hagen, John P. The Viking and the Vanguard. Vol. 4, no. 4 (Fall 1963): 435–451. ("Viking")

93. Hall, A. Rupert. Engineering and the scientific revolution. Vol. 2, no. 4 (Fall 1961): 333–341. ("Scientific Revolution")

94. Hall, A. R. The technical act. The changing technical act. Vol. 3, no. 4 (Fall 1962): 501–515. ("Technical Act")

95. Hall, R. Cargill. Early U.S. satellite proposals. Vol. 4, no. 4 (Fall 1963): 410–434. ("Satellite")

96. Hanieski, John F. The airplane as an economic variable: Aspects of technological change in aeronautics, 1903–1955. Vol. 14, no. 4 (October 1973): 535–552. ("Airplane")

97. Harris, John R., and C. Pris. The memoirs of Delaunay Deslandes. Vol. 17, no. 2 (April 1976): 201–216. ("Memoirs")

98. Harrison, Arthur P., Jr. Single-control tuning: An analysis of an innovation. Vol. 20, no. 2 (April 1979): 296–321. ("Single-Control Tuning")

99. Hartenberg, Richard S., and John Schmidt, Jr. The Egyptian drill and the origin of the crank. Vol. 10, no. 2 (April 1969): 155–165. ("Egyptian Drill")

100. Hartner, Willie. The place of humanism in a technological world. Vol. 3, no. 4 (Fall 1962): 544–553. ("Humanism")

101. Heilbroner, Robert L. Do machines make history? Vol. 8, no. 3 (July 1967): 335–345. ("Machines")

102. Heizer, Robert F. The background of Thomsen's three-age system. Vol. 3, no. 3 (Summar 1962): 259–267. ("Three-Age System")

103. Hewlett, Richard G. Pioneering on nuclear frontiers. Vol. 5, no. 4 (Fall 1964): 512–522. ("Pioneering")

104. Hewlett, R. G. Beginnings of development in nuclear technology. Vol. 17, no. 3 (July 1976): 465–478. ("Nuclear Development")

105. Hilliard, Sam B. The dynamics of power: Recent trends in mechanization on the American farm. Vol. 13, no. 1 (January 1972): 1–24. ("Farm Mechanization")

106. Hills, R. L., and A. J. Pacey. The measurement of power in early steam-driven textile mills. Vol. 13, no. 1 (January 1972): 25–43. ("Measuring Power")

107. Hoberman, Louisa Schell. Technological change in a traditional society: The case of the Desagüe in colonial Mexico. Vol. 21, no. 3 (July 1980): 386–407. ("Desagüe")

108. Hounshell, David A. Elisha Gray and the telephone: On the disadvantages of being an expert. Vol. 16, no. 2 (April 1975): 133–161. ("Telephone")

109. Howard, Robert A. Interchangeable parts reexamined: The private sector of the American arms industry on the eve of the Civil War. Vol. 19, no. 4 (October 1978): 633–649. ("Interchangeable Parts")

110. Howland, W. E. Symposium: Technology for man. The argument: Engineering education for social leadership. Vol. 10, no. 1 (January 1969): 1–10. ("Engineering Education")

111. Hughes, Thomas P. British electrical industry lag: 1882–1888. Vol. 3, no. 1 (Winter 1962): 27–44. ("British Lag")

112. Hughes, T. P. The science-technology interaction: The case of high-voltage power transmission systems. Vol. 17, no. 4 (October 1976: 646–662. ("Power Transmission")

113. Hughes, T. P. The electrification of America: The system builders. Vol. 20, no. 1 (January 1979): 124–161. ("Electrification")

114. Hughes, T. P. Emerging themes in the history of technology. Vol. 20, no. 4 (October 1979): 697–711. ("Themes")

115. Hunter, Louis C. The living past in the Appalachias of Europe: Water-mills in southern Europe. Vol. 8, no. 4 (October 1967): 446–466. ("Water Mills")

116. Hunter, Robert jF. Turnpike construction in antebellum Virginia. Vol. 4, no. 2 (Spring 1963): 177–200. ("Turnpikes")

117. Jenkins, Reese V. Technology and the market: George Eastman and the origins of mass amateur photography. Vol. 16, no. 1 (January 1975): 1–19. ("Eastman")

118. Jensen, John H., and Gerhard Rosegger. Transferring technology to a peripheral economy: The case of lower Danube transport development, 1856–1928. Vol. 19, no. 4 (October 1978): 625–702. ("Lower Danube")

119. Jeremy, David J. Innovation in American textile technology during the early 19th century. Vol. 14, no. 1 (January 1973): 40–76. ("Textiles")

120. Jevons, F. R. The interaction of science and technology today, or, is science the mother of invention? Vol. 17, no. 4 (October 1976): 729–742. ("Mother of Invention")

121. Jewett, Robert A. Structural antecedents of the I-beam, 1800–1850. Vol. 8, no. 3 (July 1967): 346–362. ("Antecedents")

122. Jewett, R. A. Solving the puzzle of the first American rail-beam. Vol. 10, no. 3 (July 1969): 371–391. ("Puzzle")

123. Jones, Daniel P. From military to civilian technology: The introduction of tear gas for civil riot control. Vol. 19, no. 2 (April 1978): 151–168. ("Tear Gas")

124. Jones, Howard Mumford. Ideas, history, technology. Vol. 1, no. 1 (Winter 1959): 20–27. ("Ideas")

125. Joravsky, David. The history of technology in Soviet Russia and Marxist doctrine. Vol. 2, no. 1 (Winter 1961): 5–10. ("Marxism")

126. Kahn, Arthur D. Every art possessed by man comes from Prometheus: The Greek tragedians and science and technology. Vol. 11, no. 2 (April 1970): 133–162. ("Greek Tragedies")

127. Kanefsky, John, and John Robey. Steam enginees in the 18th-century Britain: A quantitative assessment. Vol. 21, no. 2 (April 1980): 161–186. ("Steam Engines")

128. Keller, Alexander G. A Renaissance humanist looks at "new" inventions: The article "Horologium" in Giovanni Tortelli's *De Orthographia*. Vol 11, no. 3 (July 1970): 345–365. ("Tortelli")

129. Keller, A. G. The missing years of Jacques Besson, inventor of machines, teacher of mathematics, distiller of oils, and Huguenot pastor. Vol. 14, no. 1 (January 1973): 28–39. ("Besson")

130. Kerker, Milton. Science and the steam engine. Vol. 2, no. 4 (Fall 1961): 381–390. ("Steam Engine")

131. Kevles, Daniel J. Federal legislation for engineering experiment stations: The episode of World War I. Vol. 12, no. 2 (April 1971): 182–189. ("Engineering Stations")

132. Kevles, D. J. Scientists, the military, and the control of postwar defense research: The case of the Research Board for National Security, 1944–1946. Vol. 16, no. 1 (January 1975): 20–47. ("Postwar Research")

133. Kilgour, Frederick G. Vitruvius and the early history of wave theory. Vol. 4, no. 3 (Summer 1963): 282–286. ("Vitruvius")

134. Kohlmeyer, Fred W., and Floyd L. Herum. Science and engineering in agriculture: A historical perspective. Vol. 2, no. 4 (Fall 1961): 368–380. ("Agriculture")

135. Krammer, Arnold. Fueling the Third Reich. Vol. 19, no. 3 (July 1978): 394–422. ("Third Reich")

136. Kranzberg, Melvin. At the start. Vol. 1, no. 1 (Winter 1959): 1–10. ("At the Start")

137. Kraus, Jerome. The British electron-tube and semiconductor industry, 1935–1962. Vol. 9, no. 4 (October 1968): 544–561. ("Electron-Tube")

138. Kren, Claudia. The traveler's dial in the late Middle Ages: The chilinder. Vol. 18, no. 3 (July 1977): 419–435. ("Chilinder")

139. Kreutz, Barbara M. Mediterranean contributions to the medieval mariner's compass. Vol. 14, no. 3 (July 1973): 367–383. ("Mariner Compass")

140. La Force, J. Clayburn. Technological diffusion in the 18th century: The Spanish textile industry. Vol. 5, no. 3 (Summer 1964): 322–343. ("Spanish Textiles")

141. Layton, Edwith T., Jr. Mirror-image twins: The communities of science and technology in 19th century America. Vol. 12, no. 4 (October 1971): 562–580. ("Mirror-Image")

142. Layton, E. T., Jr. Technology as knowledge. Vol. 15, no. 1 (January 1974): 31–41. ("Knowledge")

143. Layton, E. T., Jr. American ideologies of science and engineering. Vol. 17, no. 4 (October 1976): 688–701. ("Ideologies")

144. Layton, E. T., Jr. Scientific technology, 1845–1900: The hydraulic turbine and the origins of American industrial research. Vol. 20, no. 1 (January 1979): 64–89. ("Turbine")

145. Leicester, Henry M. Chemnistry, chemical technology, and scientific progress. Vol. 2, no. 4 (Fall 1961): 352–356. ("Chemistry")

146. Leighton, Albert C. The mule as a cultural invention. Vol. 8, no. 1 (January 1967): 45–52. ("Mule")

147. Leighton, A. C. Secret communication among the Greeks and Romans. Vol. 10, no. 2 (April 1969): 139–154. ("Secret Communication")

148. Leslie, Stuart W. Charles F. Kettering and the copper-cooled engine. Vol. 20, no. 4 (October 1979): 752–776. ("Copper-Cooled Engine")

149. Lieberstein, Samuel. Technology, work and sociology in the USSR: The NOT movement. Vol. 16, no. 1 (January 1975): 48–66. ("Soviet NOT Movement")

150. Lienhard, John H. The rate of technological improvement before and after the 1830's. Vol. 20, no. 3 (July 1979): 515–530. ("Rate of Improvement")

151. Loria, Mario. Cavour and the development of the fertilizer industry in Piedmont. Vol. 8, no. 2 (April 1967): 159–177. ("Cavour")

152. Mark, Robert, John Abel, and James K. Chiu. Stress analysis of historical structures: Maillart's warehouse at Chiasso. Vol. 15, no. 1 (January 1974): 49–63. ("Stress Analysis")

153. Massouh, Michael. Innovations in street railways before electric traction: Tom L. Johnson's contributions. Vol. 18, no. 2 (April 1977): 175–201. ("Street Railways")

154. Mayo-Wells, Wilfrid J. The origins of space telemetry. Vol. 4, no. 4 (Fall 1963): 499–514. ("Telemetry")

155. Mayr, Otto. Adam Smith and the concept of the feedback system: Economic thought and technology in 18th century Britain. Vol. 12, no. 1 (January 1971): 1–23. ("Adam Smith")

156. Mayr, O. Yankee practice and engineering theory: Charles T. Porter and the dynamics of the high-speed steam engine. Vol. 16, no. 4 (October 1975): 570–602. ("Porter Engine")

157. Mayr, O. The science-technology relationship as a historiographic problem. Vol. 17, no. 4 (October 1976): 663–673. ("Historiographic Problem")

158. Mazlish, Bruce. The fourth discontinuity. Vol. 8, no. 1 (January 1967): 1–15. ("Fourth Discontinuity")

159. McCutcheon, W. A. Inland navigation in the north of Ireland. Vol. 6, no. 4 (Fall 1865): 596–620. ("Inland Navigation")

160. McLuhan, Marshall, and Barrington Nevitt. The argument: Causality in the electric world. Vol. 14, no. 1 (January 1973): 1–18. ("Causality")

161. Meeker, Joseph W. The imminent alliance: New connections among art, science, and technology. Vol. 19, no. 2 (April 1978): 187–198. ("Art")

162. Mesthene, Emmanuel G. On understanding change: The Harvard University program on technology and society. Vol. 6, no. 2 (Spring 1965): 222–235. ("Harvard")

163. Mesthene, E. G. An experiment in understanding: The Harvard program two years after. Vol. 7, no. 4 (Fall 1966): 475–492. ("Harvard Two Years After")

164. Mesthene, E. G. Some general implications of the research of the Harvard University program on technology and society. Vol. 10, no. 4 (October 1969): 489–513. ("Implications")

165. Miles, Wyndham D. The Polaris. Vol. 4, no. 4 (Fall 1963): 478–489. ("Polaris")

166. Miller, Harry. Potash from wood ashes: Frontier technology in Canada and the United States. Vol. 21, no. 2 (April 1980): 187–208. ("Potash")

167. Morris, Bertram. The context of technology. Vol. 18, no. 3 (July 1977): 395–418. ("Context")

168. Muendel, John. The horizontal mills of medieval Pistoia. Vol. 15, no. 2 (April 1974): 194–225. ("Mills")

169. Multhauf, Robert P. The scientist and the "improver" of technology. Vol. 1, no. 1 (Winter 1959): 38–47. ("Improver")

170. Multhauf, R. P. A museum case history: The Department of Science and Technology of the United States Museum of History and Technology. Vol. 6, no. 1 (Winter 1965): 47–58. ("Museum")

171. Multhauf, R. P. Sal ammoniac: A case history of industrialization. Vol 6, no. 4 (Fall 1965): 569–586. ("Sal Ammoniac")

172. Multhauf, R. P. The French crash program for saltpeter production, 1776–1794. Vol. 12, no. 2 (April 1971): 163–181. ("Saltpeter")

173. Multhauf, R. P. Observations on the state of the history of technology. Vol. 15, no. 1 (January 1974): 1–12. ("History of Technology")

174. Multhauf, R. P. Geology, chemistry, and the production of common salt. Vol. 17, no. 4 (October 1976): 634–645. ("Salt")

175. Mumford, Lewis. History: Neglected clue to technological change. Vol. 2, no. 3 (Summer 1961): 230–236. ("Neglected Clue")

176. Mumford, L. Authoritarian and democratic technics. Vol. 5, no. 1 (Winter 1964): 1–9. ("Authoritarian Technics")

177. Mumford, L. Man the finder. Vol. 6, no. 3 (Summer 1965): 375–381. ("Man the Finder")

178. Mumford, L. Technics and the nature of man. Vol. 7, no. 3 (Summer 1966): 303–317. ("Nature of Man")

179. Nelson, Alan H. Six-wheeled carts: An underview. Vol. 13, no. 3 (July 1972): 391–416. ("Carts")

180. Nicholas, S. J. The American export invasion of Britain: The case of the engineering industry, 1870–1914. Vol. 21, no. 4 (October 1980): 570–588. ("Export Invasion")

181. Nunis, Doyce B., Jr., Garland Fulton, and Charles J. McCarthy. Oral history and the history of technology. Vol. 4, no. 2 (Spring 1963): 149–176. ("Oral History")

182. Overfield, Richard A. Charles E. Bessey: The impact of the "new" botany on American agriculture, 1880–1910. Vol. 16, no. 2 (April 1975): 162–181. ("Bessey")

183. Packer, James E. Structure and design in ancient Ostia: A contribution to the study of Roman imperial architecture. Vol. 9, no. 3 (July 1968): 357–388. ("Ostia")

184. Parr, J. Gordon. The sinking of the *Ma Robert*: An excursion into mid-19th century steelmaking. Vol. 13, no. 2 (April 1972): 209–225. ("Ma Robert")

185. Paterson, Alan M. Oranges, soot, and science: The development of frost protection in California. Vol. 16, no. 3 (July 1975): 360–376. ("Oranges")

186. Pearson, Lee M. The "Princeton" and the "Peacemaker": A study in nineteenth-century naval research and development procedures. Vol 7, no. 2 (Spring 1966): 163–183. ("Peacemaker")

187. Pendray, G. E. Pioneer rocket development in the United States. Vol. 4, no. 4 (Fall 1963): 384–392. ("Rocket")

188. Perkins, John H. Reshaping technology in wartime: The effect of military goals on entomological research and insect-control practices. Vol. 19, no. 2 (April 1978): 169–186. ("DDT")

189. Perry, Robert L. The Atlas, Thor, and Titan. Vol 4, no. 4 (Fall 1963): 466-477. ("Atlas")

190. Poole, J. B., and R. Reed. The preparation of leather and parchment by the Dead Sea Scrolls Community. Vol. 3, no. 1 (Winter 1962): 1–28. ("Dead Sea")

191. Post, Robert C. The Page locomotive: Federal sponsorship of invention in mid-19th-century America. Vol. 13, no. 2 (April 1972): 140–169. ("Page Locomotive")

192. Post, R. C. "Liberalizers" versus "scientific men" in the antebellum patent office. Vol. 17, no. 1 (January 1976): 24–54. ("Patent Office")

193. Price, Derek J. DeSolla. Automata and the origins of mechanism and mechanistic philosophy. Vol. 5, no. 1 (Winter 1964): 9–24. ("Automata")

194. Price, D. J. D. Is technology historically independent of science? Vol. 6, no. 4 (Fall 1965): 553–568. ("Technology Independent")

195. Puhvel, Jaan. The Indo-European and Ido-Aryan plough: A linguistic study of technological diffusion. Vol. 5, no. 2 (Spring 1964): 176–190. ("Plough")

196. Pursell, Carroll W., Jr. Tariff and technology: The foundation and development of the American tin-plate industry, 1872–1900. Vol. 3, no. 3 (Summer 1962): 267–284. ("Tin-Plate")

197. Pursell, C. W., Jr. A preface to government support of research and development: Research legislation and the National Bureau of Standards, 1935–41. Vol. 9, no. 2 (April 1968): 145–164. ("Bureau of Standards")

198. Pursell, C. W., Jr. The Technical Society of the Pacific Coast, 1884–1914. Vol. 17, no. 4 (October 1976): 702–717. ("Technical Society")

199. Pursell, C. W., Jr. Government and technology in the great depression. Vol. 20, no. 1 (January 1979): 162–174. ("Government")

200. Rae, John B. The "know-how" tradition: Technology in American history. Vol. 1, no. 2 (Spring 1960): 139–150. ("Know-How")

201. Rae, J. B. Science and engineering in the history of aviation. Vol. 2, no. 4 (Fall 1961): 391–399. ("Aviation")

202. Rasmussen, Wayne D. Advances in American agriculture: The mechanical tomato harvester as a case study. Vol. 9, no. 4 (October 1968): 531–543. ("Tomato")

203. Reingold, Nathan. Alexander Dallas Bache: Science and technology in the American idiom. Vol. 11, no. 2 (April 1970): 163–177. ("Bache")

204. Reti, Ladislao. Francesco di Giorgio Martini's treatise on engineering and its plagiarists. Vol. 4, no. 3 (Summer 1963): 287–298. ("Martini")

205. Reti, L. Leonardo and Ramelli. Vol. 13, no. 4 (October 1972): 577–605. ("Leonardo")

206. Reynolds, Terry S. Scientific influences on technology: The case of the overshot waterwheel, 1752–1754. Vol. 20, no. 2 (April 1979): 170–295. ("Waterwheel")

207. Rezneck, Samuel. The European education of an American chemist and its influence in 19th-century America: Eben Norton Horsford. Vol. 11, no. 3 (July 1970): 366–388. ("Horsford")

208. Roach, Robert D. The first rocket-belt. Vol. 4, no. 4 (Fall 1963): 490–498. ("Rocket-Belt")

209. Robinson, Eric. James Watt and the patent law. Vol. 13, no. 2 (April 1972): 115–139. ("Watt")

210. Roland, Alex. Bushnell's submarine: American original or European import? Vol. 18, no. 2 (April 1977): 157–174. ("Bushnell")

211. Rosenberg, Nathan. Economic development and the transfer of technology: Some historical perspectives. Vol. 11, no. 4 (October 1970): 550–575. ("Transfer")

212. Rosenberg, N. Technology and the environment: An economic exploration. Vol. 12, no. 4 (October 1971): 543–561. ("Environment")

213. Rosenberg, N. Technological interdependence in the American economy. Vol. 20, no. 1 (January 1979): 25–50. ("Interdependence")

214. Rosenbloom, Richard S. Some 19th-century analyses of mechanization: Men and machines. Vol. 5, no. 4 (Fall 1964): 489–511. ("Mechanization")

215. Rürup, Reinhard. Historians and modern technology: Reflections on the development and current problems of the history of technology. Vol. 15, no. 2 (April 1974): 161–193. ("History of Technology")

216. Ruttan, V. W., and Yujiro Hayami. Technology transfer and agricultural development. Vol. 14, no. 2 (April 1973): 119–151. ("Transfer")

217. Sandler, Stanley. The emergence of the modern British capital ship. Vol. 11, no. 4 (October 1970): 576–595. ("Capital Ship")

218. Schallenberg, Richard H., and David A. Ault. Raw materials supply and technological change in the American charcoal iron industry. Vol. 18, no. 3 (July 1977): 436–466. ("Charcoal Iron")

219. Scherer, F. M. Invention and innovation in the Watt-Boulton steam-engine venture. Vol. 6, no. 2 (Spring 1965): 165–187. ("Watt-Boulton")

220. Schmandt-Besserat, Denise. The envelopes that bear the first writing. Vol. 21, no. 3 (July 1980): 357–385. ("Envelopes")

221. Sharrer, G. Terry. The indigo bonanza in South Carolina, 1740–1790. Vol. 12, no. 3 (July 1971): 447–455. ("Indigo")

222. Shelby, Lon R. Setting out the keystones of pointed arches: A note on medieval "baugeometrie." Vol. 10, no. 4 (October 1969): 537–548. ("Keystones")

223. Shriver, Donald W., Jr. Man and his machines: Four angles of vision. Vol. 13, no. 4 (October 1972): 531–555. ("Man and Machines")

224. Simms, D. L. Archimedes and the burning mirrors of Syracuse. Vol. 18, no. 1 (January 1977): 1–24. ("Archimedes")

225. Sinclar, Bruce. At the turn of a screw: William Sellers, the Franklin Institute, and a standard American thread. Vol. 10, no. 1 (January 1969): 20–34. ("Screw Thread")

226. Sinclair, B. Canadian technology: British traditions and American influences. Vol. 20, no. 1 (January 1979): 108–123. ("Canadian Technology")

227. Skolimowski, Henryk. The structure of thinking in technology. Vol. 7, no. 3 (Summer 1966): 371–383. ("Thinking")

228. Skramstad, Harold. The Georgetown canal incline. Vol. 10, no. 4 (October 1969): 549–560. ("Canal Incline")

229. Sleeswyk, Andre Wegener. The Celestial River: A reconstruction. Vol. 19, no. 3 (July 1978): 423–449. ("Celestial River")

230. Smith, Cyril Stanley. Methods of making chain mail (14th to 18th centuries): A metallographic note. Vol. 1, no. 1 (Winter 1959): 60–67. ("Chain Mail")

231. Smith, C. S. The interaction of science and practice in the history of metallurgy. Vol. 2, no. 4 (Fall 1961): 357–367. ("Metallurgy")

232. Smith, C. S. The discovery of carbon in steel. Vol 5, no. 2 (Spring 1964): 149–175. ("Carbon")

233. Smith, C. S. The cover design: Art, technology, and science: Notes on their historical interaction. Vol. 11, no. 4 (October 1970): 493–549. ("Art")

234. Smith, Merritt Roe. John H. Hall, Simeon North, and the nature of technological innovation among the antebellum arms makers. Vol. 14, no. 4 (October 1973): 573–591. ("Hall and North")

235. Smith, Thomas M. Project Whirlwind: An unorthodox dvelopment project. Vol. 17, no. 3 (July 1976): 447–464. ("Whirlwind")

236. Solo, Robert A. The meaning and measure of economic progress. Vol. 9, no. 3 (July 1968): 389–414. ("Economic Progress")

237. Spence, Clark C. Early uses of electricity in American agriculture. Vol. 3, no. 2 (Spring 1962): 142–160. ("Electricity")

238. Spong, Raymond A. The Boyers' locomotive toy. Vol. 13, no. 3 (July 1972): 373–390. ("Toy")

239. Strassman, W. Paul. The risks of innovation in twentieth century manufacturing methods. Vol. 5, no. 2 (Spring 1964): 215–223. ("Risks")

240. Strauss, Felix F. "Mills without wheels" in the 16th century Alps. Vol. 12, no. 1 (January 1971): 23–43. ("Mills")

241. Sun, E-Tu Zen. Wu Ch'i-Chün: Profile of a Chinese scholar-technologist. Vol. 6, no. 3 (Summer 1965): 394–406. ("Chinese Scholar")

242. Susskind, Charles, and Arlene Inouye. "Technological Trends and National Policy," 1937: The first modern technology assessment. Vol. 18, no. 4 (October 1977): 593–621. ("Technology Assessment")

243. Tascher, John. U.S. rocket society number two: The story of the Cleveland Rocket Society. Vol. 7, no. 1 (Winter 1966): 48–63. ("Cleveland")

244. TeBrake, William H. Air pollution and fuel crises in pre-industrial London, 1250–1650. Vol. 16, no. 3 (July 1975): 337–359. ("Pollution")

245. Theobald, Robert. Emerging nations: Long-term prospects and problems. Vol. 3, no. 4 (Fall 1962): 601–616. ("Emerging Nations")

246. Thomas, Donald E., Jr. Diesel, father and son: Social philosophies of technology. Vol. 19, no. 3 (July 1978): 376–393. ("Diesel")

247. Tobey, Ronald. Theoretical science and technology in American ecology. Vol. 17, no. 4 (October 1976): 718–728. ("Ecology")

248. Tokaty, G. A. Soviet rocket technology. Vol. 4, no. 4 (Fall 1963): 515–528. ("Soviet Rockets")

249. Tucker, D. Gordon. François van Rysselberghe: Pioneer of long-distance telephony. Vol. 19, no. 4 (October 1978): 650–674. ("Long-Distance Telephony")

250. Uselding, Paul. Elisha K. Root, forging and the "American System." Vol. 15, no. 4 (October 1974): 543–568. ("Forging")

251. Usher, Abbott Payson. The industrialization of modern Britain. Vol. 1, no. 2 (Spring 1960): 109–127. ("Industrialization")

252. Vanek, Joann. Household technology and social status: Rising living standards and status and residence differences in housework. Vol. 19, no. 3 (July 1978): 361–375. ("Housework")

253. Vincenti, Walter G. The air-propeller tests of W. F. Durand and E. P. Leslie: A case study in technological methodology. Vol. 20, no. 4 (October 1979): 712–751. ("Tests")

254. Virginsky, V. S. The birth of steam navigation in Russia and Robert Fulton. Vol. 9, no. 4 (October 1968): 562–569. ("Steam Navigation")

255. Vogel, Robert M. Assembling a new hall of civil engineering. Vol. 6, no. 1 (Winter 1965): 59–73. ("Civil Engineering")

256. Wachsmann, Klaus P., and Russell Kay. African ethnomusicology: the interrelations of instruments, music forms, and cultural systems. Vol. 12, no. 3 (July 1971): 399–413. ("Musical Instruments")

257. Watson-Watt, Sir Robert. Technology in the modern world. Vol. 3, no. 4 (Fall 1962): 385–393. ("Modern World")

258. Welsh, Peter C. A craft that resisted change: American tanning practices to 1850. Vol. 4, no. 3 (Summer 1963): 299–317. ("Tanning")

259. Welsh, P. C. An American contribution to hand-tool design: The metallic woodworking plane. Vol. 7, no. 1 (Winter 1966): 38–47. ("Wood Plane")

260. White, Lynn, Jr. Eilmer of Malmesbury, an eleventh century aviator. Vol. 2, no. 2 (Spring 1961): 97–111. ("Malmesbury")

261. White, L. The act of invention: Causes, contexts, continuities, and consequences. Vol. 3, no. 4 (Fall 1962): 486–500. ("Invention")

262. White, L. The study of medieval technology, 1924–1974: Personal reflections. Vol. 16, no. 4 (October 1975): 519–530. ("Reflections")

263. Wik, Reynold Millard. Henry Ford's science and technology for rural America. Vol. 3, no. 3 (Summer 1962): 247–258. ("Ford")

264. Wik, R. M. Benjamin Holt and the invention of the track-type tractor. Vol. 20, no. 1 (January 1979): 90–107. ("Tractor")

265. Wilkinson, Norman B. Brandywine borrowings from European technology. Vol. 4, no. 1 (Winter 1963): 1–13. ("Brandywine")

266. Wilkinson, N. B. In anticipation of F. W. Taylor: A study of work by Lammot du Pont, 1872. Vol. 6, no. 2 (Spring 1965): 208–221. ("Anticipating Taylor")

267. Wise, George. A new role for professional scientists in industry: Industrial research at General Electric, 1900–1916. Vol. 21, no. 3 (July 1980): 408–429. ("GE Research")

268. Woodbury, Robert S. The legend of Eli Whitney and interchangeable parts. Vol. 1, no. 3 (Summer 1960): 235–253. ("Whitney Myth")

269. Woodruff, Helga, and William Woodruff. Economic growth: Myth or reality? The interrelatedness of continents and the diffusion of technology, 1860–1960. Vol. 7, no. 4 (Fall 1966): 453–474. ("Diffusion")

270. Znachko-Iavorskii, Igor L. New methods for the study and contemporary aspects of the history of cementing materials. Vol. 18, no. 1 (January 1977): 25–42. ("Cement")

271. Zvorikine, A. The history of technology as a science and as a branch of learning: A Soviet view. Vol. 2, no. 1 (Winter 1961): 1–4. ("Soviet History")

272. Zvorikine, A. Technology and the laws of its development. Vol. 3, no. 4 (Fall 1962): 443–458. ("Technology")

Books and Dissertations (with short titles)

Aitken, Hugh G. J. *Taylorism at the Watertown Arsenal.* Cambridge, Mass.: Harvard University Press, 1960. (*Taylorism*)

Aitken, Hugh G. J. *Syntony and Spark: The Origins of Radio.* New York: John Wiley & Sons, 1976. (*Syntony*)

Allison, David Kite. *New Eye for the Navy: The Origins of Radar at the Naval Research Laboratory.* Washington, D.C.: Naval Research Laboratory (NRL Report 8466), 1981. (*New Eye*)

Almond, Gabriel A., Marvin Chodorow, and Roy Harvey Pearce, eds. *Progress and Its Discontents.* Berkeley: University of California Press, 1982. (*Progress and Its Discontents*)

Berger, Peter L., and Thomas Luckmann. *The Social Construction of Reality: A Treatise in the Sociology of Knowledge.* Garden City, N.Y.: Doubleday & Co., Anchor Books, 1966. (*Social Construction*)

Billington, David P. *Robert Maillart's Bridges.* Princeton, N.J.: Princeton University Press, 1979. (*Maillart*)

Bodner, John. *Immigration and Industrialization: Ethnicity in an American Mill Town, 1870–1940.* Pittsburgh: University of Pittsburgh Press, 1977. (*Immigration and Industrialization*)

Boorstin, Daniel J. *The Americans: The Democratic Experience.* New York: Random House, 1973. (*Americans*)

Braverman, Harry. *Labor and Monopoly Capital: The Degradation of Work in the Twentieth Century.* New York: Monthly Review Press, 1974. (*Labor and Monopoly Capital*)

Brody, David. *Steelworkers in America: The Non-Union Era.* New York: Harper and Row, 1960. (*Steelworkers*)

Buder, Stanley. *Pullman: An Experiment in Industrial Order and Community Planning, 1890–1930*. New York: Oxford University Press, 1967. (*Pullman*)

Bulliet, Richard W. *The Camel and the Wheel*. Cambridge, Mass.: Harvard University Press, 1975. (*Camel*)

Burlingame, Roger. *March of the Iron Men: A Social History of Union through Invention*. New York: Grosset & Dunlap, The Universal Library, 1938. (*Iron Men*)

Butterfield, Herbert. *The Whig Interpretation of History*. London: G. Bell and Sons, 1951. (*Whig Interpretation*)

Byington, Margaret. *Homestead: The Households of a Mill Town*. New York: Arno, 1969. (*Homestead*)

Cardwell, Donald S. L. *From Watt to Clausius: The Rise of Thermodynamics in the Early Industrial Age*. Ithaca, N.Y.: Cornell University Press, 1971. (*Clausius*)

Cebik, L. B. *Concepts, Events, and History*. Washington, D.C.: University Press of America, 1978. (*Concepts*)

Chandler, Alfred D., Jr. *The Visible Hand: The Managerial Revolution in American Business*. Cambridge, Mass.: Belknap Press of Harvard University Press, 1977. (*Visible Hand*)

Collins, James. *Descartes' Philosophy of Nature*. Oxford: American Philosophical Quarterly Monograph Series, 1971. (*Descartes*)

Constant, Edward W., II. *The Origins of the Turbojet Revolution*. Baltimore: Johns Hopkins University Press, 1980. (*Turbojet*)

Cowan, Ruth Schwartz. *More Work for Mother: The Ironies of Household Technology from the Open Hearth to the Microwave*. New York: Basic Books, 1983. (*More Work for Mother*)

Cumbler, John T. *Working-Class Community in Industrial America: Work, Leisure, and Struggle in Two Industrial Cities, 1880–1930*. Westport, Conn.: Greenwood Press, 1979. (*Working-Class Community*)

Daumas, Maurice, ed. *A History of Technology and Invention: Progress through the Ages*. 3 vols. Trans. by Eileen B. Hennessy. New York: Crown Publishers, 1969–1979. (*History of Technology*)

Daumas, Maurice, ed. *Histoire Generale des Techniques*. Paris: Presses Universitaires de France, 1962–1968. (*Histoire*)

Dawley, Alan. *Class and Community: The Industrial Revolution in Lynn*. Cambridge, Mass.: Harvard University Press, 1976. (*Class and Community*)

D'Onofrio-Flores, Pamela M., and Sheila M. Pfafflin, eds. *Scientific-Technological Change and the Role of Women in Development*. Boulder, Colo.: Westview Press, 1982. (*Role of Women*)

Dublin, Thomas. *Women at Work: The Transformation of Work and Community in Lowell, Mass., 1826–1860*. New York: Columbia University Press, 1979. (*Women at Work*)

Durbin, Paul T., ed. *A Guide to the Culture of Science, Technology and Medicine*. New York: Free Press, 1979. (*Science, Technology, Medicine*)

Edwards, Richard C. *Contested Terrain: The Transformation of the Work Place in the Twentieth Century*. New York: Basic Books, 1979. (*Contested Terrain*)

Finch, James Kip. *The Story of Engineering*. Garden City, N.Y.: Doubleday & Co., Anchor Books, 1960. (*Engineering*)

Foster, John. *Class Struggle and the Industrial Revolution: Early Industrial Capitalism in Three English Towns*. London: Weidenfeld and Nicolson, 1974. (*Class Struggle*)

Freire, Paulo. *Pedagogy of the Oppressed.* Trans. by Myra B. Ramos. New York: Herder and Herder, 1970. (*Pedagogy of the Oppressed*)

Gadamer, Hangs-Georg. *Truth and Method.* New York: Seabury Press, 1975. Trans. by Garrett Barden and John Cumming. Originally published as *Wahrheit und Methode.* Tübingen: J. C. B. Muhr, 1960. (*Truth and Method*)

Gilfillan, S. Collum. *Inventing the Ship.* Chicago: Follett Publishing Co., 1935. (*Inventing the Ship*)

Gilfillan, S. C. *The Sociology of Invention: An Essay in the Social Causes, Ways and Effects of Technic Invention, Especially as Demonstrated Historically in the Author's "Inventing the Ship."* Cambridge, Mass.: MIT Press, 1970. Reissue of the original 1935 edition. (*Sociology of Invention*)

Gilligan, Carol. *In a Different Voice: Psychological Theory and Women's Development.* Cambridge, Mass.: Harvard University Press, 1982. (*Different Voice*)

Goodenough, Ward Hunt. *Cooperation in Change.* New York: Russell Sage Foundation, 1963. (*Change*)

Gutman, Herbert. *Work, Culture, and Society in Industrializing America.* New York: Random House, 1966. (*Work, Culture, and Society*

Harding, Sandra, and Merril B. Hintikka, eds. *Discovering Reality.* D. Reidel Publishing Co., 1983. (*Discovering Reality*)

Hareven, Tarmara K., and Randolph Langenbach. *Amoskeag: Life and Work in an American Factory-City.* New York: Pantheon Books, 1978. (*Amoskeag*)

Hatfield, H. Stafford. *The Inventor and His World.* New York: E. P. Dutton and Co., 1933. (*Inventor*)

Hindle, Brooke. *Emulation and Invention.* New York: New York University Press, 1982. (*Emulation*)

Hirsch, Susan E. *Roots of the American Working Class: The Industrialization of Crafts in Newark, 1800–1860.* Philadelphia: University of Pennsylvania Press, 1978. (*Crafts in Newark*)

Holley, I. B., Jr. *Ideas and Weapons: Exploitation of the Aerial Weapon by the United States during World War One.* New Haven, Conn.: Archon, 1953. (*Ideas and Weapons*)

Holsti, Ole R. *Content Analysis for the Social Sciences and Humanities.* Reading, Mass.: Addison-Wesley Publishing Co., 1969. (*Content Analysis*)

Hounshell, David A. *From the American System to Mass Production, 1800–1932: The Development of Manufacturing Technology in the United States.* Baltimore: Johns Hopkins University Press, 1984. (*American System*)

Hounshell, D. A., ed. *The History of American Technology: Exhilaration or Discontent?* Wilmington, Del.: Hagley Papers, 1984. (*Exhilaration or Discontent*)

Hughes, Thomas Parke. *Elmer Sperry: Inventor and Engineer.* Baltimore: Johns Hopkins University Press, 1971. (*Sperry*)

Hughes, T. P. *Networks of Power: Electrification in Western Society, 1880–1930.* Baltimore: Johns Hopkins University Press, 1983. (*Networks*)

Hunter, Louis. *Steamboats on the Western Rivers: An Economic and Technological History.* Cambridge, Mass.: Harvard University Press, 1949. (*Steamboats*)

Hunter, L. *Waterpower in the Century of the Steam Engine.* University Press of Virginia for the Eleutherian Mills-Hagley Foundation, 1980. (*Waterpower*)

Jacques Cattell Press, eds. *Directory of American Scholars*. 4 vols. New York: R. R. Bowker Co., 1978. 7th ed. (*American Scholars*)

al-Jazari, Ism'il ibn al-Razzaz. *The Book of Knowledge of Ingenious Mechanical Devices*. Trans. and annotated by Donald R. Hill. Boston: D. Reidel Publishing Co., 1973. (*Ingenious Mechanical Devices*)

Jenkins, Reese V. *Images and Enterprise: Technology and the American Photographic Industry, 1829–1925*. Baltimore: Johns Hopkins University Press, 1975. (*Images and Enterprise*)

Jeremy, David J. *Transatlantic Industrial Revolution: The Diffusion of Textile Technologies between Britain and America, 1790–1830*. Cambridge, Mass.: MIT Press and Merrimack Valley Textile Museum, 1981. (*Textile Technologies*)

Jones, Gareth Stedman. *Outcast London: A Study in the Relationship between Classes in Victorian Society*. London: Clarendon Press of Oxford University Press, 1971. (*Outcast London*)

Josephson, Matthew. *Edison: A Biography*. New York: McGraw-Hill, 1957. (*Edison*)

Kasson, John F. *Civilizing the Machine: Technology and Republican Values in America, 1776–1900*. New York: Grossman Publishers, 1976. (*Civilizing the Machine*)

Keller, Evelyn Fox. *A Feeling for the Organism: The Life and Work of Barbara McClintock*. San Francisco: W. H. Freeman and Co., 1983. (*McClintock*)

Keller, E. F. *Reflections on Gender and Science*. New Haven, Conn.: Yale University Press, forthcoming. (*Gender*)

Keyser, Conrad. *Bellifortis*. 2 vols. Trans. and ed. by Götz Quarg. Düsseldorf: Verlag des Vereins Deutscher Ingenieure, 1967. (*Bellifortis*)

Klem, Friedrich. *A History of Western Technology*. Cambridge, Mass.: MIT Press, 1964. Originally published as *Technik: Eine Geschichte ihrer Probleme*. Freiburg: Karl Alber, 1954. (*History of Technology*)

Koppes, Clayton R. *JPL and the American Space Program: A History of the Jet Propulsion Laboratory*. New Haven, Conn.: Yale University Press, 1982. (*JPL*)

Kranzberg, Melvin, and Carroll W. Pursell, Jr., eds. *Technology in Western Civilization*. 2 vols. New York: Oxford University Press, 1967. (*Technology*)

Lasker, G. E., ed. *Applied Systems and Cybernetics*. Vol. 2: *Systems Concepts, Models, and Methodology*. New York: Pergamon Press, 1981. (*Applied Systems and Cybernetics*)

Laudan, Larry. *Progress and Its Problems: Towards a Theory of Scientific Growth*. Berkeley: University of California Press, 1977. (*Progress and Its Problems*)

Layton, Edwin T., Jr. *The Revolt of the Engineers: Social Responsibility and the American Engineering Profession*. Cleveland: Case Western Reserve University Press, 1971. (*Revolt*)

Leibniz, Gottfried Wilhelm. *The Monadology of Leibniz*. Trans. and ed. by Herbert W. Carr. Los Angeles: University of Southern California Press, 1930. (*Monadology*)

Mack, Pamela E. *The Politics of Technological Change: A History of Landsat*. Ph.D. dissertation, University of Pennsylvania. Ann Arbor, Mich.: University Microfilms, 1983. (*Landsat*)

May, Rollo. *The Courage to Create*. New York: W. W. Norton & Co., 1975. (*Courage*)

Merchant, Carolyn. *The Death of Nature: Women, Ecology, and the Scientific Revolution*. New York: Harper and Row, 1980. (*Death of Nature*)

Meyer, Stephen. *The Five Dollar Day: Labor, Management and Social Control in the Ford Motor Company, 1908–1919*. Albany: State University of New York Press, 1981. (*Five Dollar Day*)

Montgomery, David. *Workers' Control in America: Studies in the History of Work, Technology, and Labor Struggles.* Cambridge: Cambridge University Press, 1979. (*Workers' Control in America*)

Mumford, Lewis. *Technics and Civilization.* New York: Harcourt Brace & World, 1963, with new introduction; originally published in 1934. (*Technics*)

Murphey, Murray G. *Our Knowledge of the Historical Past.* New York: Bobbs-Merrill Co., 1973. (*Historical Past*)

Nelson, Daniel. *Managers and Workers: Origins of the New Factory System of the United States, 1880–1920.* Madison: University of Wisconsin Press, 1975. (*Managers and Workers*)

Nisbet, Robert. *History of the Idea of Progress.* New York: Basic Books, 1970. (*Idea of Progress*)

Noble, David F. *America by Design: Science, Technology, and the Rise of Corporate Capitalism.* New York: Alfred A. Knopf, 1979. (*America by Design*)

Noble, D. F. *Force of Production.* New York: Alfred A. Knopf, 1983. (*Forces of Production*)

Ogburn, William Fielding. *Social Change with Respect to Culture and Original Nature.* New York: Huebsch, 1923. (*Social Change*)

Perrin, Noel. *Giving Up the Gun: Japan's Reversion to the Sword, 1543–1879.* Boulder, Colo.: Shambhala, 1980. (*Giving Up the Gun*)

Polanyi, Michael. *Personal Knowledge: Towards a Post-Critical Philosophy.* Chicago: University of Chicago Press, 1958. (*Personal Knowledge*)

Prude, Jonathan. *The Coming of Industrial Order: Town and Factory Life in Rural Massachusetts, 1810–1860.* New York: Cambridge University Press, 1983. (*Industrial Order*)

Roland, Alex. *Model Research: The National Advisory Committee for Aeronautics, 1915–1958.* Washington, D.C.: NASA (SP-4103), 1984. (*Model Research*)

Rothschild, Joan, ed. *Machina ex Dea: Feminist Perspectives on Technology.* New York: Pergamon Press, 1983. (*Machina ex Dea*)

Schmookler, Jacob. *Invention and Economic Growth.* Cambridge, Mass.: Harvard University Press, 1966. (*Invention*)

Schon, Donald A. *Displacement of Concepts.* London: Tavistock Publications, 1963. (*Displacement of Concepts*)

Schon, D. A. *Technology and Change: The New Heraclitus.* New York: Delacorte Press, 1967. (*Technology and Change*)

Schumpeter, Joseph Alois. *The Theory of Economic Development: An Inquiry into Profits, Capital, Credit, Interest, and the Business Cycle.* Cambridge, Mass.: Harvard University Press, 1934. (*Economic Development*)

Schumpeter, J. A. *Business Cycles: A Theoretical, Historical and Statistical Analysis of the Capitalist Process.* 2 vols. New York: McGraw-Hill, 1939. (*Business Cycles*)

Shaiken, Harley. *Automation and Work in the Computer Age.* New York: Holt, Reinhart and Winston, 1983. (*Automation*)

Sinclair, Bruce. *Philadephia's Philosopher Mechanics: A History of the Franklin Institute, 1824–1865.* Baltimore: Johns Hopkins University Press, 1974. (*Franklin Institute*)

Singer, Charles, E. J. Holmyard, A. R. Hall, and Trevor I. Williams, eds. (assisted by E. Jaffe, R. G. H. Thomson, and J. M. Donaldson). *A History of Technology.* 5 vols. London: Oxford University Press, 1954–1958. (*History of Technology*)

Sivin, Nathan, ed. *Science and Technology in East Asia.* New York: Science History Publications, 1977. (*Science and Technology in East Asia*)

Smith, Adam. *Works.* Ed. by Dugald Stewart. 5 vols. London: Cadell, 1811. Vol. 5: *Principles Which Lead and Direct Philosophical Enquiries, Illustrated by the History of Astronomy.* (*Principles*)

Smith, Cyril Stanley, ed. *Sources for the History of the Science of Steel, 1532–1786.* Cambridge, Mass.: MIT Press and The Society for the History of Technology, 1968. (*Steel*)

Smith, Merritt Roe. *Harpers Ferry Armory and the New Technology: The Challenge of Change.* Ithaca, N.Y.: Cornell University Press, 1977. (*Harpers Ferry*)

Smith, M. R., ed. *Military Enterprise and Technological Change: Perspectives on the American Experience.* Cambridge, Mass.: MIT Press, forthcoming. (*Military Enterprise*)

Spiegel-Rösing, Ina, and Derek De Solla Price, eds. *Science, Technology, and Society: A Cross-Disciplinary Perspective.* Beverly Hills, Calif.: Sage Publications, 1977. (*Science, Technology and Society*)

Spradley, James P., ed. *Culture and Cognition: Rules, Maps, and Plans.* San Francisco: Chandler Publishing Co., 1972. (*Culture and Cognition*)

Staudenmaier, John M., S.J. *Design and Ambience: Historians and Technology, 1958–1977.* Ph.D. dissertation, University of Pennsylvania. Ann Arbor, Mich.: University Microfilms, 1980. (*Design and Ambience*)

Stearns, Peter N., and Daniel Walkowitz, eds. *Workers in the Industrial Revolution: Recent Studies of Labor in the United States and Europe.* New Brunswick, N.J.: Rutgers University Press, 1974. (*Workers in the Industrial Revolution*)

Taylor, George R. *The Transportation Revolution: 1815–1860.* Armonk, N.Y.: M. E. Sharpe, 1951. (*Transportation Revolution*)

Thompson, E. P. *The Making of the English Working Class.* New York: Pantheon Books, 1964. (*English Working Class*)

Trescott, Martha Moore, ed. *Dynamos and Virgins Revisited: Women and Technological Change in History.* Metuchen, N.J.: Scarecrow Press, 1979. (*Dynamos and Virgins*)

Usher, Abbott Payson. *A History of Mechanical Inventions.* Cambridge, Mass.: Harvard University Press, 1929. (*Mechanical Inventions*)

Walkowitz, Daniel J. *Worker City, Company Town: Iron and Cotton-Worker Protest in Troy and Cohoes, New York, 1855–1884.* Urbana: University of Illinois Press, 1978. (*Worker City, Company Town*)

Wallace, Anthony F. C. *Culture and Personality.* New York: Random House, 1961. (*Culture and Personality*)

Wallace, A. F. C. *Rockdale: The Growth of an American Village in the Early Industrial Revolution.* New York: Alfred A. Knopf, 1978. (*Rockdale*)

Weber, Michael. *Social Change in an Industrial Town: Patterns of Progress in Warren, Pennsylvania, from the Civil War to World War I.* University Park: Pennsylvania State University Press, 1976. (*Warren, Pennsylvania*)

White, Lynn, Jr. *Medieval Technology and Social Change.* New York: Oxford University Press, 1962. (*Medieval Technology*)

White, L., Jr. *Machina ex Deo: Essays in the Dynamism of Western Culture.* Cambridge, Mass.: MIT Press, 1968. (*Machina ex Deo*)

Winner, Langdon. *Autonomous Technology: Technics-out-of-Control as a Theme in Politcal Thought.* Cambridge, Mass.: MIT Press, 1977. (*Autonomous Technology*)

Wulff, Hans Eberhard. *The Traditional Crafts of Persia: Their Development, Technology, and Influence on Eastern and Western Civilization.* Cambridge, Mass.: MIT Press, 1967. (*Crafts of Persia*)

Zinberg, Dorothy S., ed. *Uncertain Power: The Struggle for a National Energy Policy.* New York: Pergamon Press, 1983. (*Uncertain Power*)

Zvorikine, A., et al., eds. *Geschichte der Technik.* 2 vols. Leipzig, 1964. Originally published in the U.S.S.R., 1962. (*Geschichte*)

Articles and Unpublished Papers (with short titles)

Bruner, Jerome S., Jacqueline J. Goodnow, and George A. Austin. Categories and cognition. In Spradley, ed., *Culture and Cognition.* ("Categories and Cognition")

Bush, Corlann Gee. Women and the assessment of technology: To think, to be, to unthink, to free. In Rothschild, ed., *Machina ex Dea.* ("Women")

Donovan, Arthur. Coal industry in Appalachia. Paper read at 22nd annual meeting of SHOT, Newark, N.J., October 1979. ("Coal Industry")

Dunlavy, Colleen A. Transcending internalism: On the historiography of technology in the United States. Unpublished manuscript written for the Program in Science, Technology and Society at MIT, February 1983. ("Transcending Internalism")

Frake, Charles O. The ethnographic study of cognitive systems. In Spradley, ed., *Culture and Cognition.* ("Cognitive Systems")

Gearhart, Sally M. An end to technology: A modest proposal. In Rothschild, ed., *Machina ex Dea.* ("Modest Proposal")

Goodwin, Jack. Current bibliography in the history of technology. TC 6, no. 1 (Spring 1965): 346–74; annually thereafter. ("Current Bibliography")

Hafter, Daryl. Craftwomen in pre-industrial Europe. Paper read at 22nd annual meeting of SHOT, Newark, N.J., October 1979. ("Craftwomen")

Hage, Per. Münchner beer categories. In Spradley, ed., *Culture and Cognition.* ("Beer Categories")

Hall, A. Rupert. "A History of Technology": Some editorial reflections. TC 1, no. 3 (Fall 1960): 311–319. ("History of Technology")

Hall, A. R. More on medieval pivoted axles. TC 2 (Winter 1961): 17–22. ("Axles")

Hindle, Brooke. "The exhilaration of early American technology": A new look. In Hounshell, ed., *Exhilaration or Discontent?*, pp. 7–17. ("Exhilaration")

Hoberg, George. Determinism vs possibility: On technology, social science, and society. Unpublished manuscript for the MIT Department of Political Science, September 1983. ("Determinism")

Hounshell, David A. Commentary: On the discipline of the history of American technology. *The Journal of American History* 67 (March 1981): 854–865. ("History of American Technology")

Hounshell, D. A. Letter to the editor. *The Journal of American History* 68 (1981–1982): 900–902. ("Letter")

Hughes, Thomas Parke. Introduction: The development phase of technological change. TC 17, no. 3 (July 1976): 423–431. ("Development")

Hughes, T. P. Technological momentum in history: Hydrogenation in Germany, 1893–1933. *Past and Present* 441 (August 1969): 106–132. ("Technological Momentum")

Keller, Evelyn Fox. Baconian science: A hermaphroditic birth. *The Philosophical Forum* 11, no. 3 (Spring 1980): 299–308. ("Baconian Science")

Keller, E. F., and Christine Grontkowski. The mind's eye. In Harding and Hintikka, eds., *Discovering Reality*. ("Mind's Eye")

Keller, E. F. Women, science, and popular mythology. In Rothschild, ed., *Machina ex Dea*. ("Women, Science")

Keohane, Nannerl O. The enlightenment idea of progress revisited. In Almond, Chodorow, and Pearce, eds., *Progress and Its Discontents*. ("Enlightenment Idea of Progress")

King, Ynestra. Toward an ecological feminism and a feminist ecology. In Rothschild, ed., *Machina ex Dea*. ("Ecological Feminism")

Larson, J. Lauritz. A systems approach to the history of technology: An American railroad example. Unpublished manuscript. Joan Calihin Robinson Prize-winning paper read at the 1980 annual meeting of SHOT, Toronto. ("Systems Approach")

Leslie, Stuart. Exhibit review. The urban habitat: The city and beyond. TC 23, no. 3 (July 1982): 417–429. ("Urban Habitat")

McGaw, Judith. Machines and women's work: The case of nineteenth century America. Paper read at 22nd annual meeting of SHOT, Newark, N.J., October 1979. ("Machines and Women's Work")

MacLeod, Roy. Changing perspectives in the social history of science. In Spiegel-Rösing and Price, eds., *Science, Technology, and Society*. ("Changing Perspectives")

Merchant, Carolyn. Mining the earth's womb. In Rothschild, ed., *Machina ex Dea*. ("Mining the Earth's Womb")

Noble, David. In defense of Luddism. *Democracy* TC 3, nos. 2, 3, 4 (Spring, Summer, Fall 1983). ("Luddites")

Oldroyd, D. R. Sir Archibald Geikie (1835–1924), geologist, romantic aesthete, and historian of geology: The problem of Whig historiography of science. *Annals of Science* 37 (1980): 441–462. ("Whig Historiography")

Peterson, James. Technology and work in the twentieth century: A bibliographical overview. Paper read at 22nd annual meeting of SHOT, Newark, N.J., October 1979. ("Technology and Work")

Porter, Glenn. Commentary. A historiographical revolution: A. F. C. Wallace's *Rockdale* and Merritt Roe Smith's Harpers Ferry. Paper read at 22nd annual meeting of SHOT, Newark, N.J., October 1979. ("Historiographical Revolution")

Price, Derek J. De Solla. On the historiographic revolution in the history of technology: Commentary on the papers by Multhauf, Ferguson, and Layton. TC 15, no. 1 (January 1974): 42–48. ("Commentary")

Reingold, Nathan, and Arthur Molella. Introduction: The interaction of science and technology in the industrial age. Proceedings of the Burndy Library Conference. TC 17, no. 4 (October 1976): 624–633. ("Science and Technology")

Rothschild, Joan. Introduction: Why *Machina ex Dea?* In Rothschild, ed., *Machina ex Dea.* ("Introduction")

Schmookler, Jacob. The changing efficiency of the American economy, 1869–1938. *Review of Economics and Statistics* 34 (1952): 214–231. ("Changing Efficiency")

Schmookler, J. Patent application statistics as an index of inventive activity. *Journal of the Patent Office Society* 35 (1953): 539–550. ("Patent Statistics")

Schmookler, J. The level of inventive activity. *Review of Economics and Statistics* 36 (1954): 183–190. ("Inventive Activity")

Selove, Richard. Energy policy and democratic theory. In Zinberg, ed., *Uncertain Powers.* ("Energy Policy")

Smith, Merritt Roe. The "American System" of manufacturing. In Smith, M. R., ed., *Military Enterprise.* ("American System")

Stapleton, Darwin H. Letter to the editor. In *Journal of American History* 68 (1980–1981): 897–900. ("Letter")

Staudenmaier, John M., S.J. What SHOT hath wrought and what SHOT hath not: Reflections on twenty-five years of the history of technology. TC 25, no. 4 (October 1984). ("What SHOT Hath Wrought")

Thackray, Arnold. The history of science. In Durbin, ed., *Science, Technology, Medicine.* ("History of Science")

Thackray, A. Review of *Philadelphia's Philosopher Mechanics: A History of the Franklin Institute, 1824–1865*, Bruce Sinclair. TC 17, no. 2 (April 1976): 383–384. ("Franklin Institute: Review")

Thompson, E. P. Time, work-discipline, and industrial capitalism. *Past and Present* 38 (1967): 56–97. ("Time")

Tozer, Lowell. A century of progress, 1833–1933: Technology's triumph over man. *American Quarterly* 4 (Spring 1952): 206–209. ("Century of Progress")

Wallace, Anthony F. C. Culture and cognition. In Spradley, ed., *Culture and Cognition.* ("Culture and Cognition")

Zuboff, Shoshanah. Work and human interaction in historical perspective. Cambridge, Mass.: Harvard University, 1979. Unpublished manuscript for the Harvard Business School. ("Work and Human Interaction")

Zwick, Martin. Dialectical thermodynamics. In Lasker, ed., *Applied Systems and Cybernetics.* ("Dialectical Thermodynamics")

Index